T0345902

MINNESOTA’S GEOLOGIST

Also by Sue Leaf
Published by the University of Minnesota Press

A Love Affair with Birds: The Life of Thomas Sadler Roberts

The Bullhead Queen: A Year on Pioneer Lake

Portage: A Family, a Canoe, and the Search for the Good Life

Minnesota's GEOLOGIST

The Life *of* Newton Horace Winchell

SUE LEAF

University of Minnesota Press
Minneapolis • London

The publication of this book was assisted by a bequest from Josiah H. Chase
to honor his parents, Ellen Rankin Chase and Josiah Hook Chase,
Minnesota territorial pioneers.

Published by the University of Minnesota Press
111 Third Avenue South, Suite 290
Minneapolis, MN 55401-2520
http://www.upress.umn.edu

ISBN 978-1-5179-0168-4 (hc)

A Cataloging-in-Publication record for this book
is available from the Library of Congress.

Printed in the United States of America on acid-free paper

25 24 23 22 21 20   10 9 8 7 6 5 4 3 2 1

The world is the geologist's great puzzle box; he stands before it like the child to whom the separate pieces of his puzzle remain a mystery til he detects their relation and sees where they fit, and then his fragments grow at once into a connected puzzle beneath his hand.

—LOUIS AGASSIZ, *GEOLOGICAL SKETCHES* (1866)

CONTENTS

INTRODUCTION

THERE IS SUCH A THING AS A GEOLOGICAL EYE. Some people can scan a landscape and see glacial moraines where others see hills, or detect ancient seabeds in what look to most of us like cliffs.

For years, I had walked blind to this. But not my husband. Long ago, he and I were driving back roads east of St. Paul, looking for migrating ducks, when he sighed, "I just love this glaciated terrain. These kettle lakes are textbook perfect."

Glacial terrain? Kettle lakes? I'd been looking for hooded mergansers.

Newton Winchell had a geological eye. Traveling into the Red River valley for the first time, he saw a prehistoric lake bed where glacial meltwater had pooled. Sailing along Lake Superior's north shore, he saw cooled magma, the remains of volcanoes. As the head of the Minnesota Geological and Natural History Survey from 1872 until 1900, it was his job to see geologically, to discern what lay beneath the surface, whether stone for building or accessible groundwater, and how it could be wisely used.

Geological eyes are not inherited or randomly bestowed on a person. They are a result of training, of study and reflection. The eyes are sharpened by exposure to basalt outcrops, eroded river valleys, and upended rock layers that show how the earth has tossed itself about. They are honed by sifting through glacial till, dirtying hands, picking up stones, and scrutinizing cleavage planes.

At twenty, Newton Winchell declared, "I know nothing about rocks." At twenty-five, he had decided he would make them his life's work. This is the story of a poor boy struggling to overcome a family background of idealism run amok, who emerged from that family with a high regard for education, a fine methodical mind, and a penchant for hard work, vowing he would never be poor again.

Geology never made Winchell rich, precisely because his inclination toward idealism pushed him to pursue perfection in whatever he did. His ethical code prevented him from making money on the rocks he knew would provide great wealth to others; there is no hint in any of his papers that he considered investing in mining claims in the iron districts he examined and mapped.

But Winchell was more than a geologist to Minnesota. He was an early feminist, married to the first woman elected to office in Minnesota. He was an affectionate father to a brood of children who each left their mark on the state. He was a Methodist who walked the tightrope between science and religion, he was a historian who anchored his major work in context, and he became an archaeologist who wrote an early definitive work on the Native Americans of the place now called Minnesota.

Newton Horace Winchell was a keystone scientific figure in Minnesota history. Young men attuned to nature, who liked to tromp around outdoors, were drawn to him because of what he knew and who he was. Amateurs of science (and there were many in the 1800s) were likewise attracted, because in his teacherly guise he strove to make geology accessible to the nonprofessional. Bluff and hearty, with a big laugh and a bushy beard, Winchell knew intimately nearly every corner of the state. At a time when few roads crisscrossed Minnesota, he traveled by train, horse, and cart, by canoe and sailing boat, looking at river valleys, railroad cuts, rock quarries, and well bores in his attempt to piece together the underlying geology of the state.

One of the charms of elucidating the life of Newton Horace Winchell has been that geology abides. Towns may grow bigger or fade away, roads might be cut and widened, and farm fields may disappear to suburban tracts: what humans build is subject to change. But the limestone cliffs at Winona, the bluffs at Red Wing, the pipestone quarry in southwestern Minnesota, and Palisade Head on the North Shore—these look mostly the same as they did when Winchell cast his geological eye on them. Rocks are bigger than human endeavor. In an age when human influence seeps into the most remote corners of the planet, it is oddly pleasing that rocks endure.

Understanding the geology of the place we call home fosters an appreciation of how old the earth is, how much it has changed, and how dynamic and restless it really is. Geology challenges humans to accept that we are not the center of all things meaningful, not the center of anything at all. Reality is bigger and wilder than we have imagined. Newton Winchell knew this, and lived by it when most educated people had only just begun to awaken to the meaning of geology. As you follow him on these pages, I hope that you gain fresh appreciation for our untamed planet. I hope you, too, will see with geological eyes.

1

FINDING A FOOTING

1839–57

NEWTON WINCHELL BENT OVER HIS JOURNAL, barely able to decipher the lines as he recorded the events of his day. August crickets chirped in the dark, and the candle flame flickered, casting big shadows on the wall as a gentle wisp of air came through the window.

He ran his fingers through a shock of dark hair, bleached by the summer sun, and sighed. Another day spent pursuing what he called "the root of all evil," another day of struggle to help his mother keep the wolf from the door.

Had he glanced in a mirror, he would have seen a callow young man with large, luminous eyes, a broad brow, and an expression on his face that was both wistful and hopeful. His childhood had not been carefree. Ensconced in a warm, loving family, he had lost an older sister to an early death, watched his father become increasingly immersed in arcane polemics, and his mother scramble to care for their brood.

His older siblings—he had three—had encouraged Newton to keep an account of his life that summer, and so he had plunked down good money for the leather-bound journal. "I bought this book intending to note down such fortunes and misfortunes as I consider worthy of notice," he wrote. "I might at some future day be interested in perusing what befell me in my younger days."[1]

The summer of 1855 was not going as planned. Newton had wanted to attend school. At fifteen, he had completed eight grades and aimed for higher education. He read and wrote and calculated. He studied history and geography. But he felt the need to brush up on grammar, and his grasp of Latin was shaky. He passed his evenings reading a book by Fowler on phrenology, the study of the conformation of the human skull, used in predicting mental capacity and character. This "science" was all the rage, and Newton found it interesting as well as useful.[2]

There were other amusements in the evening. The Winchell family was musical. Both parents had fine voices, as did their children.[3] The family had a piano, and his sisters played. Newton often joined in what he termed "little sings" before bedtime, a way of socializing that filled a notoriously meager house with harmony.

The family was short of money that summer. Finances were always strained in the Winchell family, with seven children, but funds were especially tight right then, and Newton's parents thought he should spend his summer working for neighboring farmers, contributing to the family income.

But that wasn't going as planned, either. Rain and wind had hampered haying and the small grain harvest. Newton spent wet days at home, running errands for his mother or studying, "which I think is quite as useful as working to me," he observed.[4] Still, by mid-August, he had earned about six dollars.

The Winchell family, Horace and Caroline and their youngest children, were living in Lakeville, in northwestern Connecticut, snug against the New York State and Massachusetts borders. They had left North East,[5] New York, four miles to the west of Lakeville, because their little hamlet, Spencer's Corner, was losing population to its bigger neighbor, Millerton, which had been awarded the prize of the railroad line in 1851. Opportunities dwindled for Caroline's millinery business.

In moving, Newton's family left behind a slew of relatives—siblings of Horace, Caroline's mother, cousins, and extended family, a web of relations that had put down roots deep into Winchell Mountain, which rose over North East and was named for the family.

Lakeville was "a very pleasant village," Newton agreed, but he was homesick for North East, where the Winchell family had resided for almost a century.[6] That homestead, Newton confided in his journal, was "where I have enjoyed more pleasure than in any place I ever lived. The place where I have become very much attached, the place I once called home."[7]

So Newton was most unsettled when his parents proposed later in August to sell their homestead in North East in order to pay off accumulating debt. His mother, the family bookkeeper, had trouble collecting what was owed to the family. Boarders, when she took on boarders, often didn't pay or didn't pay in full. Caroline couldn't cover the bills. Everything Newton had earned went to household expenses. Occasionally, his father also brought in money working on local farms,[8] but Horace spent most of his time reading and writing impassioned tracts on various topics. He seemed indifferent to the domestic concern of family expenses.[9] These were considerable. In addition to Newton, there were three other children at home: Antoinette, age thirteen; Robby, age eleven; and Charles, age nine.

It was Caroline who drove a rented team of horses to North East to investigate selling the homestead—the family was too poor to own a horse and wagon—Caroline who undertook the restoration of their current home in Lakeville, whitewashing, papering, and laying carpet; and Caroline who eventually took in boarders to make ends meet.[10] Caroline also ran her milliner's shop within the home, crafting hats and other items, which required her to take annual trips to New York City for supplies.[11] Later, Caroline devised another income-generating scheme by teaching a "numerical method of dress cutting" in surrounding communities,[12] netting about a dollar a day. Caroline McAllister Winchell was Newton's first experience of a strong, competent, intelligent woman, but she would not be the only take-charge woman in his life.

Meanwhile, her husband closeted himself in scholarly pursuit.[13] Reflective and well-read, Horace had been educated for the ministry. His grandfather had been an early convert to the Baptist church in America, and his older brother, James Manning Winchell, was the celebrated pastor at a prestigious Baptist church in Boston. But early on, Horace became dissatisfied with Baptist dogma and plowed his own theological path. His handwritten manuscripts contained references to the early Christian theologian Tertullian, Melanchthon's Augsburg Confession, and Rabbi Moses Maimonides, the Jewish philosopher. They mentioned Thomas Burnet, who opined about Noah's flood, and Bishop Ussher, who calculated the age of the earth (6,000 years), based on the generations of people mentioned in the Bible, beginning with Adam. Horace's writings also had quotes from Edward

Gibbon's *Decline and Fall of the Roman Empire* and John Bunyan's *Pilgrim's Progress,* with reference to first-century Jewish historian Josephus thrown in.[14]

Most of Horace's writing concerned societal reforms. One intense focus was on the abolishment of church denominations and the unification of all Christians, a cause an adult Newton would cast in positive terms as "sacred and impelling."[15] Another favorite cause decried the sin of slavery, and a third preached the destruction of the wicked.[16] "Father thinks the end of the time is near," Newton recorded in his journal, "and consequently is out warning the people."[17]

When Caroline was gone from home "selling her charts," Antoinette and Robby boarded with older sister Laura, married and living in Lenox, Massachusetts, and Charley, the youngest child, went to live in North East with Grandma McAllister. Horace remained at home, reliant on townspeople for handouts. Newton was mortified by his father's failure to provide for his family. "He seems to have no self respect or shame at all," he wrote to his brother.[18]

The family's two older sons, Alexander and Martin, were already out of the house and embarked on careers, and Newton aimed to be like them. "I want to get off somewhere like you and Alex," he told his brother Martin, "where I am a stranger and my antecedents are not known."[19] Alexander, married and a father at age thirty-two, had recently assumed the chair of geology, zoology, and botany at the University of Michigan at Ann Arbor; Martin, twenty-nine, was living in New Jersey and practicing medicine. Both men remained in contact with the family, advising their mother on money matters,[20] counseling their younger brothers and sister, and arranging visits. Alexander was in an excellent position to further Newton's education and urged him to come to Ann Arbor, board with him, and go to school.[21] He thought that Newton could earn his board by helping around the household before and after school; tuition itself would cost Newton ten dollars.

Newton deemed the plan workable, and it interested him.[22] He was desperate to distance himself from his family. "Domestic misfortune and family disturbance. I am sick of [it] entirely," he told Martin. "If I can get away for a short time I am glad."[23] The big problem would be finding money for the train ticket to Ann Arbor, and for decent clothes to wear when attending school.

By the end of August 1855, Newton had decided he would apply for a teaching position in a local school. He did not need state certification but would be examined for competency by the local board hiring him.

School teaching was in his very blood. His parents had both taught;[24] Martin had taught before becoming a physician; his younger sister, Antoinette, was preparing to teach; and of course, Alex, whom he admired greatly, had had a diversity of teaching posts, from seminaries to academies, teaching girls, boys, and now college students. At fifteen, Newton was young for a teacher, and he decided he should start in a small school.[25]

Two weeks passed between Newton's writing this decision in his journal and his actual pursuit of a position. Life was full of distractions. His mother's boarders came and went. A church picnic was diverting. Traveling Methodists set up camp in a woods near town and commenced revival meetings. Interested, Newton investigated the worshipful gatherings. "Sinners are flocking to their Maker," he reported."[26] Scores of fallen souls were redeemed, but Newton didn't record being among the favored.

When the Methodists had taken down their tent and moved on, Newton set out on foot to apply for positions. He applied to schools close to home. He applied to schools farther away. Fall rains made the roads muddy and impassable at times, but he doggedly pursued his goal. "I declare I *must* have *one,*" he wrote after a particularly wet spell had kept him near home. "Moreover, I must have some money for some clothes."[27] He had great hope of a position at a school in his native North East—how wonderful to be in his hometown once again!—but then got word that they wanted to hire a woman.[28]

Finally, something came through. The school on Winchell Mountain needed a schoolmaster, and even better, his father's youngest brother, George, was in charge of hiring. When asked, Newton told the trustees of the school that he wanted twelve dollars a month. The committee told him they could pay him only ten dollars a month, to which Newton replied, "I'd rather do something else than teach for less."[29]

However, when he related the story that night at home, Martin, visiting from New Jersey, told him to rethink the matter. Teaching for ten dollars a month was preferable to not teaching at all. After reflection, Newton agreed, and the next Monday walked the six miles back to the mountain and took the job.[30]

The school at Winchell Mountain was a public school serving local children in about a two-square-mile area,[31] who walked to school. The school was overseen by a committee of trustees, elected annually; Uncle George Winchell was one of these. Public schools received some state funds, but local residents were responsible for more than three-fourths of the cost of maintaining the schoolhouse and the teacher.[32]

Newton was told to expect twelve scholars. The first day, he had two. The next day, no children showed up, and he closed the school until the next week.[33] The term would last until mid-March, and Newton was determined to remain for the duration. He was very aware of his young age and had in those opening weeks one student who was older than he. "I mistrust [him]," Newton confided in his journal. "Thinks he is going to rule the school and teacher from appearances but I pray he will be vastly mistaken."[34]

Newton taught all eight grades in one room at Winchell Mountain. The school was heated by a wood stove, which the trustees had agreed to supply with wood. As winter settled in and the stove burned wood more rapidly, Newton arrived at school in the morning several days to face an empty wood box. It was not Newton's job to split wood, and he thought he might have to suspend classes, but he no sooner entertained the thought when two of his oldest scholars appeared with axes in hand.

Like most teachers of the times, Newton boarded with local families as part of his pay. Having his room and meals paid for allowed him to save his entire weekly pay, and boarding gave him firsthand experience with his pupils' home life.

District teachers maintained a home base at the trustee's residence. They kept their trunks there and could return to it on weekends.[35] Newton's home base was undoubtedly his Uncle George's farm, which had been the original family homestead where his father, Horace, had been raised. The homestead perched at the summit of Winchell Mountain, a fertile plateau that the family had cultivated for three generations.

From the farm, the views were splendid. Looking west, the distant Catskills hugged the horizon. Looking east, the blue hills of western Connecticut marked where Newton's parents now lived. Back on the mountain named for his family, Newton had distanced himself from the financial and domestic chaos that haunted his folks. He focused on his work and saved for his future.

Winter term ended the third week in March, and though Newton knew the trustees were anxious to have him teach through the summer of 1856, no contract was forthcoming.[36] Newton once more resumed the job hunt, walking in his spare hours from town to town. Most schools had already hired for summer term, and he got tired and discouraged. It didn't help his mood that his younger sister was also applying to teach and had very favorable prospects.[37] He considered other ways of making money.

"Money! That is all I stand in need of now particularly," he wrote in frustration. "With that I can accomplish any ends and plans that I can apply as an aid to my designs viz. *attending school.* I attain to something else besides school *teaching.*"[38]

With no prospects, Newton decided to establish his own school in a room in the Lakeville home. He had no regrets about leaving his former classroom behind—the families there still owed him thirteen dollars.[39]

Newton opened his own school in the family home for summer term. It flourished, but family financial woes continued. Horace sought publication of one of his apocalyptic tracts; Caroline continued to trim hats and feed boarders. The renters remained behind on their payments. Debts accumulated. Summer turned to fall. As winter approached, everyone, including Newton, believed that they would need to sell their place in order to meet the mortgage on it.[40]

Amid this chaotic and uncertain atmosphere, several signs promised a way out: Alex renewed his invitation for Newton to come west to Michigan; the school committee at Millerton, New York, four miles west of Lakeville, hired Newton to teach winter term at the princely sum of twenty dollars per month; and a group of Lakeville men organized a lyceum aimed at improving public speaking. Newton was one of twenty organizers and was designated to speak at the very first meeting.

The trustees at Millerton gave Newton a one-month trial period and then guaranteed him a position through the winter. He taught thirty-five students, some of them older and bigger than he was, and he taught all subjects, including algebra, philosophy, and history.

The job of a schoolteacher in a one-room school demanded a broad-based knowledge, a mastery of multitasking, and the ability to control many children of a wide range of ages. A typical day began at nine o'clock with roll call, followed by a short Bible reading. Classes began at nine fifteen and rotated at fifteen-minute intervals. Each grade level was called to the front to recite to the teacher, while the other grades worked at their desks. More time was devoted to learning to read than all other subjects.[41]

In addition to reading, children learned spelling, penmanship, grammar, and pronunciation using phonics. A popular spelling text of the time used fables imparting a moral lesson at the end, illustrated by woodcuts. Enhancing these areas of English, Newton also had his students working on declamation—recitation—and weekly compositions.[42]

Classes recessed at midmorning for fifteen minutes in the schoolyard and then continued until noon, when students ate lunch. If it was too far to walk home for the midday meal, students brought lunches in pails, which were kept on shelves above the coats at the back of the room. A typical school would have a half-hour lunch break.[43]

In addition to the "two R's" of reading and writing, Newton taught "'rithmetic": five levels of it, and higher math, algebra. He also taught three different geography classes. There was a second recess in midafternoon, and school let out around three thirty, after which the frazzled teacher prepped, if need be, for the next day.

To reinforce spelling skills, Newton appealed to his students' competitive instincts and often held bees during Saturday classes. "I let the spellers choose sides and combat each other . . . until one or the other is vanquished," he wrote in his journal, recapping his day. As a public school, it had "scholars of all dispositions and abilities." Newton observed, "I have a good share of rough characters to deal with, and I have others also of finer and more polished manners."[44]

Newton boarded at his Grandma McAllister's at Spencer's Corner, and after Christmas, his youngest brother, Charley, joined him. By January 1857, the Winchell family in Lakeville had scattered—Robby and Antoinette going to live with Laura and her husband in Lenox, and Caroline on the road in search of clients. Letters between the siblings went back and forth.[45]

Horace, too, may have been away from home that winter. He was still trying to publish his lengthy tome warning of the End. He collected subscriptions from would-be readers to cover publishing costs. Newton observed that his father "trusts Providence for a maintenance,"[46] but told Martin that their father in truth expected charity from the neighbors.[47] Newton shrank in shame.

Although he had felt at the start of Millerton's winter term that he now had enough experience in the classroom to succeed in everything, there were always those boys who were bigger and older than he was, sitting at the back of the room. One day in January, in the depths of winter's snow and cold, Newton had a run-in with one of them.

Nicholas Best is a large fellow who has a little picture—possibly a girlie sketch. It distracts him from his studies. Newton asks him to put it away, and Nicholas does—temporarily. But it soon reappears, and finally, Newton requests the picture.

When Nicholas refuses, Newton reaches for the small whip that he keeps at his desk. He is not used to being defied in class, and his heart is pounding. He is being put to the test.

When Nicholas refuses one last request, Newton strikes him with the whip. Nicholas tries to snatch the whip, and Newton strikes again and again. Then, as Newton's students watch goggle-eyed, Nicholas rises from his desk, seizes the whip from his teacher's hand and sits down.

A prolonged scuffle ensues with Newton roaring, "Do you intend to mind me?" and Nicholas yelling, "No!" At last, Newton regains the whip and the upper hand. Collecting his wits, he asks one last time, "Do you intend to mind me?" This time, Nicholas answers, "yes," but Newton tells him to gather his books and leave. "We want no boy here that will not mind," Newton says. "If I was hired to fight I should fight, but I was not. . . . Boys that will not obey without fighting must be somewhere besides our school."

Incredibly, the incident is not over. Nicholas, in taking up his books, can't find his pen. Newton helps him hunt for the pen, and as both teens search together, they have a little time to reflect.

"Well," says Newton, to bring the incident to a close, "I will promise to go to the store and get you all the pens you want, . . . [but] we are waiting and want to get rid of you as soon as

possible for it is getting late. I will get you the pens tonight and call on you."

"But," says Nicholas. "I want to come to school."

Newton thinks quickly. "No," he says. "We cannot have you here. You can't mind me."

Nicholas repeats that he wants to be in school. He will obey his teacher.

Newton looks at the student and gauges his sincerity. "I am perfectly willing that you should attend my school," he says, "as long as you treat me as your teacher and obey me unconditionally and without hesitation." Nevertheless, Nicholas needs to be disciplined, because he deserves it. It could be considered a re-admittance fee.

Nicholas steps forward, and Newton then says, "I do not want to hurt you. Since you have set a good example and came here of your own accord, I shall not strike you but three blows." And lastly, he asks for the little picture that started it all, and Nicholas hands it over.

"There now," Newton said. "We are good friends again. We shall forget this, at least I shall, but we shall always remember that the scholar must be submissive to the teacher." Newton shook hands with his student and then turned to the class on the platform that had been waiting to recite, standing at attention through the entire fracas.

THE MILLERTON SCHOOL BOARD closed winter term early. Attendance dropped off, and Newton's twenty dollars per month pay seemed outsized to some of the trustees. He had saved a lot of money during the term—ninety dollars, more than he had ever had in his life—but he also had debts to pay off. Still, he had enough to get to Ann Arbor to his brother Alexander's place. Martin urged him strongly to go.[48]

Newton had given some thought to staying in Lakeville and helping his mother financially, but his older brothers saw the whole family situation as a losing venture. They advised their mother to sell the Lakeville house, and their younger brother to pursue his own life.[49] It was time for the ambitious teen to leave.

Newton had one final task with the Millerton school—collecting the tuition due him. He was afraid that he would never see the money, since he was leaving. On a rainy day, he doggedly walked from family to family, asking for his pay. Grandma McAllister lent him an old blue umbrella, which proved useless. He became soaked to the skin. After twelve hours, he dragged himself back home, dark hair plastered to his head, almost too tired to peel off his wet clothes. "I was numb, cold, tired and wet," he later recalled, "but *I had my money* that was what I was *bound* for *that day*."[50]

The next morning, Newton Winchell boarded the train for the two-day journey to Ann Arbor. Alexander had written him two pages of specific instructions on how to get to Michigan. Newton carried an umbrella in his hand the entire way, more to see if he could keep track of it than having need of it—a talisman of sorts. His younger brother Robby bet that he would get lost, but Robby lost that wager. On April 18, 1857, at age seventeen, Newton arrived at Alexander's house, ready to start a new life.

A SOLID FOUNDATION

1857–64

NEWTON ARRIVED IN ANN ARBOR with $9.56 in his pocket.[1] It was all he had left of his "princely" salary in Millerton. He had bought necessary clothes, and the train fare depleted his savings. It cost ten dollars to enter the University of Michigan. He didn't know where that money would come from—something to worry about later. First, he needed to settle in at Alexander's house, strike a deal to cover room and board, and learn if he was adequately prepared for college work.

In 1857, Ann Arbor, Michigan, was a thriving college town of five thousand people. The university boasted a picturesque campus shaded by oaks and consisted of Greek Revival buildings with triangular pediments and solid columns. Mason Hall, built in 1840, had been the first academic building. University Hall, its mirror image, stood to its south. Across campus, medical students were educated in the three-storied Medical Building. A chemistry building had recently been added.[2]

Newton liked Ann Arbor. He liked the climate, at least in the cool, tender days of April. He liked his brother's family: his sister-in-law, Julia, and his two little nieces, especially Julie, age five, who took a particular shine to him. He liked being useful; within days of arrival, he helped the family move to a new house and whitewash the home's interior. He liked his accommodations (a large room, all to himself) and he liked having access to Alexander's large library, which was weighted toward geology, zoology, mineralogy, botany, and agriculture.[3]

Alexander Winchell had been appointed professor of physics and civil engineering at the University of Michigan in 1853 at age twenty-nine. By 1857, his duties had shifted to a newly instituted chair of natural history.[4] Trained at Wesleyan University in Connecticut, Alexander had an aptitude for mathematics, which he also taught. He had prior experience with botany, astronomy, geology, and meteorology. In Alabama, he did field work on fossils with the Alabama Geological Survey.[5] Later, after studying the geology of oil fields, he would propose that oil becomes trapped in upfoldings, or anticlines, of rock. Oil prospectors in the future would use this insight to locate deposits.[6]

Newton hoped to model himself after his brother and relied on his judgment. Alexander tested Newton's college readiness and told him he had two options: he could enter the university under the Scientific Course in the fall, or he could attend high school for a year to make up deficiencies and enter college in the Classical Course. The Scientific Course was more technical; the Classical Course focused on Latin and Greek, modern languages, mathematics, and natural history and science. Deeming it more useful in getting a job after graduation, Newton chose the latter.

Newton threw himself into his new life. He studied Latin grammar in his spare time. His brother's family attended the Methodist–Episcopal church in town. Newton joined the choir. Ann Arbor had a debating society. Newton showed up to hear the members consider whether ministers might legitimately engage in politics (and thus preach from the pulpit against slavery; the society voted yes); whether the Roman Catholic church had been more productive of evil than good (it had, they decided); and whether the Ottoman Empire ought to be extinguished by other European powers for the benefit of mankind (yes).[7]

One Friday evening, Newton went to the university expecting to hear another debate, and learned the society had gone to see Millard Fillmore, the former president. Excitement was palpable. Newton buoyantly followed the crowd to a brass band, which marched to where Fillmore was staying. Fillmore stepped out into the chilly June night in a heavy coat, top hat in hand, and addressed the cheering townspeople. Newton found his remarks "practiced and appropriate," adding, "He is the first of the presidents I have been fortunate to see."[8]

THE POVERTY THAT HAD DOGGED NEWTON back East followed him to Michigan and would continue for the next seven years. Alexander's household served as a safety net, but Newton was essentially on his own, ever short of money, ever scrambling for funds. Once again, he turned to teaching for income. Rather than attend high school that first year in Ann Arbor, he landed a teaching position in a large school on the outskirts of town in fall 1857. He triumphed over several college students who had also applied for the job, then fretted that he didn't know enough to teach in a city school.[9]

To compensate for missing high school, he studied on his own, plodding along in Latin grammar, reading the works of Caesar, and adding Greek. He hoped he could pass the college entrance exams. He got scant support from home. His father wrote advising him to study Bible history and sacred geography if he was going to study anything. These would be useful when Christ came again.[10] Newton had no comment on that and passed his teacher qualifying exams with flying colors. Still, he mourned the fact that he had so little time for his own learning. "I sometimes wish I had lots of money," he wrote wistfully.[11]

The news from home was both cheering and distressing Newton's first year away. His closest sibling, Antoinette, had decided to become a music teacher. She confided to her brother that she intended to earn a college degree of her own.[12] But news of his older sister, Laura, was tragic. Married while still a teen, she had been a second mother to Newton. Laura died in childbirth at age twenty-six, just as Newton's teaching term was winding up. The surviving baby girl would be raised by Caroline Winchell, the baby's grandmother.[13]

After a year of teaching and a summer term attending public high school, Newton entered the University of Michigan as a freshman in 1858. He thought his command of Greek was weak,[14] but even so, he passed all entry exams. The sweaty summer hours spent laboring over Xenophon's *Anabasis* and wrestling with irregular Greek verbs paid off.[15]

Newton went into debt—again—paying college entrance fees and then went further into hock by joining a fraternity, Delta Kappa Epsilon. He chose Delta Kappa Epsilon because the best and most prominent students at the university were members, and he liked the members he knew.[16] Like most secret societies, Delta Kappa Epsilon served as a means to network and might prove influential in a career.

To attain the necessary cash, he borrowed from Alexander, and he wrote to Grandma McAllister asking for twenty-five dollars. Then he wrote to Uncle George on Winchell Mountain, and he wrote to his bachelor uncle, Lewis Winchell. He was clear in his appeals that the needed funds would accrue interest at 7 percent.[17] Grandma came through with ten dollars, which was not enough but sufficient to pay off the bookstore and the frat initiation fee. Newton worried, "I don't know exactly how I shall get along without [borrowing] money."[18] He planned to stay in school the entire year without working, because his studies demanded his full attention. He hoped to teach the following summer and, perhaps, winter.

Alexander, being a close relative, was always the last creditor Newton paid. When he tallied his debt to his brother at the end of the school year, he found he owed Alexander sixty-eight dollars, a considerable sum. Still, Newton fumed to his journal that the debt had been calculated unfairly. Now that he had seen other rentals, he deemed his room to be *very small* and overpriced. Board (provided by his sister-in-law) could also be gotten cheaper elsewhere in Ann Arbor and, besides, was not very good, especially when Alexander was away from home. He also had to do his own laundry and cleaning. He decided not to complain to Alexander, but he thought he needed a little distance from him.[19]

A college friend, Aaron Watkins, hearing of Newton's financial woes, invited him that summer to his family's farm. He could work on the farm to pay for his board and hire out to neighboring farms. In addition, the two planned to study Greek prose together.

Newton arrived at the Watkins farm in Grass Lake, Michigan, in July.[20] He was immediately put to work cutting hay and when haying was completed, cultivating corn. He would soon be needed to thresh wheat. He liked the Watkins family at once, and they him. He deemed them "nice folks for farmers"—not as fashionable as Ann Arbor residents, a relief for Newton.[21] In addition to his friend, Aaron, there was Aaron's sixteen-year-old sister, Nettie, and a little brother. Aaron and Newton worked out a daily schedule where one studied in the morning and the other drove the horses for farm work; after noon dinner, they switched roles. Town boy Newton expected that at some point there would be a lull in farm work, when both could study, but that never happened.[22]

At summer's end, the schoolmaster position at Grass Lake came open. Delaying his sophomore year of college, Newton applied for the job and got it. It was a large school with forty-five pupils on average, most of them under the age of fourteen, but also some older teens who were preparing to teach or to go on to university the next year. Newton rose to the situation and offered classes in Latin and natural philosophy (forerunner to natural science). He taught trigonometry to the older students, and phonics to the beginners as well as all the standard classes he had taught at Millerton. He organized a debate club, which was a wild success. In conjunction, he launched *The Grass Lake Review,* a magazine that printed original pieces from members of the debate club. He submitted his own poetry.[23]

Newton's intellectual energy at school was matched by the enthusiasm and support of the Grass Lake community. He had no significant discipline problems the entire term. "I believe I never had warmer or truer friends in any place, not only among the scholars but also outside the school," he confided to his journal.[24] One of his education methods books had urged young teachers to "wak[e] the community to an interest in the school district."[25] He believed he had woken up Grass Lake.

The last day of school, exceptionally warm with snowbanks melting under the March sun, Newton took all fifty students down to a nearby lake to enjoy the fine weather. They collected shells along the beach and relished the sunshine. The teenaged teacher and his students lobbed snowballs at each other. At the end, the young man stood in the door of the schoolhouse with tears in his eyes and watched his students depart. Then he gathered up his books and papers and the globe that he had used for geography lessons, a memento given to him by the school's director, and left.[26]

But Newton's ties with Grass Lake were not totally severed. He had personally tutored the daughter of the Watkins family, Nettie, who hoped to teach. She had been his best student, smart, diligent, and eager. They began exchanging letters and then portraits, and within the year they became engaged to be married. They were very young—Newton was twenty-one, and she, seventeen—but Newton assured her, "I don't intend to hurry matters. Two or three years will make a great difference to us both."[27] Nettie planned to attend the Wesleyan Seminary and Female College at Albion, Michigan, forerunner of Albion College, the following year. "She needs education," Newton wrote in his journal, "and *must* have it."[28]

Newton returned to Alexander's home after his happy time in Grass Lake. Alexander had recently moved his family to a unique house he had built himself on the edge of campus. Orson Fowler's book *The Octagon House: A Home for All* had been the inspiration for the house, which had eight sides, two stories, and a cupola. It had already become notable in Ann Arbor.[29] In spring 1860, Alex was preparing for another field season of the Michigan Geological Survey, which he headed as state geologist.[30] The survey had spent one field season in southern Michigan; this one would explore the northern shore of the Lower Peninsula. Alex hired Newton to be the botanist for the survey, though Newton had zero experience in plant identification.

Armed with a copy of Asa Gray's recently published *First Lessons in Botany and Vegetable Physiology*,[31] Newton became acquainted with the world of cotyledons, pistils, and stamens. Initially, it was a slog. He first examined marsh marigolds, spring cress, and hepatica blooming in Ann Arbor. He supplemented his study by occasionally attending the botany class on campus.[32] Slowly, he worked through the identification of thirty-four plants. Alex told him that when he had identified fifty, he would be able to pass the final examination in botany at the university, which was a junior-level course.[33] By the time the survey's team left for the north, Newton's tally was up to fifty-seven.[34]

Tragedy seemed to be stalking the Winchell family. Sharp on the heels of his sister Laura's death, a baby niece, Stella, had died of scarlet fever in November 1859. In May 1860, as the Winchell brothers prepared for the survey's expedition, Alexander's oldest daughter, Julie, Newton's favorite niece, age eight, was stricken with rheumatic fever and died.

The family was devastated by the loss of a second child to illness in less than six months. Newton poured out his grief in his journal, recounting how Julie had often preferred him to her father, how bright and curious she was, how frail and transient life was.[35] The Alexander Winchells had one surviving child, Jennie. She would be joined by a baby brother, Julian, in the coming year, but he, too, would die, of diphtheria, before turning two.[36]

Reeling from the stunning losses, Alexander nonetheless made plans to get the survey into the field that summer. The six-man team planned to travel by Mackinaw boat, a simple sailing craft popular on the upper Great Lakes in the nineteenth century. Alex went to

Mackinac Island to buy a boat, which he sent to the east side of Saginaw Bay of Lake Huron, the expedition's starting point.[37] Newton remained in Ann Arbor, working on plant identification and making arrangements for his sister Antoinette to come to Michigan for the summer and the following school year. Despite his own financial straits, he and Alexander paid for her transportation, but, alas, Newton would be in the field with the survey when she arrived.[38]

THE MICHIGAN GEOLOGICAL SURVEY left Ann Arbor for fieldwork on June 11, 1860, as the forests waxed green and the days grew long. When they reached the boat on Saginaw Bay, they hired two voyageurs to man the boat and manage the camp.[39]

The expedition's first camp was at the mouth of the Saginaw River. Loggers were decimating the white pines there. Newton counted twenty-one sawmills along a three-mile stretch of the river. The party heard there were fifty mills within twenty-one miles of the river's mouth.

The very first day, the men encountered what would become a continuing source of irritation: unfavorable winds. Alexander Winchell chose to travel by Mackinaw boat because sails could be hoisted to propel. Sails saved the crew untold time and effort, but lacking a keel, a Mackinaw struggled in headwinds.[40]

Newton's job was to collect and identify plants. He began immediately. By day two, he had identified forty-seven species, pressing others to preserve and identify later. His pay was thirty dollars a month, the same as the voyageurs'.

He brought along a small pocket journal bound in smooth calfskin in which to record the expedition, and a trunk, which housed a heavy overcoat, several changes of clothing, a sewing kit for making repairs, and his plant book. He also took a chess box to pass idle hours in camp.

The party surveyed the eastern shore of Saginaw Bay, the western side of the "thumb" of lower Michigan's "mitten." In their boat they then followed the curve of the bay north toward Alpena. Newton had a steep learning curve in geology, admitting to his journal that he "[knew] nothing about the kinds of rocks and cannot say what they are."[41] The next day, however, he identified a ledge of interest as limestone and recorded the presence of fossils.[42]

The survey was Newton's first exposure to fieldwork and to camping. The party was housed in two heavy canvas tents, often kept open at night to catch a breeze. There was no screen to keep out insects, but each man had netting to wind about the face and head. Each also had two blankets that served as a bedroll. Three weeks into the expedition, Newton began complaining about mosquitoes, joining a distinguished list of early explorers, from Lewis and Clark to Henry Schoolcraft, who found the pesky bloodsuckers insufferable.

Newton Winchell sinks to the grass and unlaces his boots. It has been a long day, and he's bushed. The survey team was up before dawn, broke camp, and set sail on the clear waters of Lake Huron. He'd had a slight headache, abnormal for him, but he couldn't be seen as a mollycoddle. He hadn't complained.

Under a beating sun, the crew examined outcrops, collected specimens, sketched maps, and took notes. Newton's head pounded, and he longed for shade and rest. He needed sleep.

The camp that night is on a gravelly beach of Huron south of Alpena. Dark conifers rise inland; grass fringes the shore. The men had pitched only one of the tents because the gravel hardpan had been impervious to tent stakes. They'd draped the second tent over the fossils collected that day.

Exhausted, Newton turns in early. Mosquitoes appear as dusk gathers, and Newton dons his old felt hat and a heavy coat. He thinks the night will turn cold. As a final flourish, he wraps mosquito netting around his head and face, and settles in. Soon the others join him, and now four grown men lie cheek by jowl in the tent, flaps left open to catch a breeze. The wind dies down. Newton feels uncomfortably warm.

Mosquitoes magnify in number. Thousands, tens of thousands, millions! Newton switches out his winter coat for a linen one. He shifts on his blanket, swatting futilely at the pests. The netting is worthless.

In the stifling tent, he wants only some sleep. But the mosquitoes!

Zing! An assault on his left earlobe. Newton swats. Fresh recruits rise to take the place of the slain.

Under bombardment, Newton hatches a plan. There are small fishing boats staked in the shallows. He'll take one and anchor offshore. The night is breathless. The lake is flat. On the water, he'll be unmolested.

Out in the boat, Newton relaxes his lanky form and adjusts to the rhythmic rocking of the water. Mosquitoes are gone, but now he finds the boat lacks an anchor, and so, he rows. As he rows, he whistles and then starts to sing.

In a clear baritone, he sings everything he can think of, folk tunes, rounds, and hymns. He sings a sentimental ditty that gives voice to his plight, out alone on the dark lake, far from Winchell Mountain:

> *Do they miss me at home, do they miss me?*
> *'Twould be an assurance most dear,*
> *To know that this moment some loved one*
> *Were saying, "I wish you were here."*

It seems he rows for hours. He tries napping, but the boat drifts. When he awakes, he is disoriented. This makes him anxious. He resumes whistling.

Finally, it seems time to go back to camp.

He pivots the boat with his oars, and now a soft breeze blows in his favor. The faintest pink edges the eastern horizon over Lake Huron. A dog barks from shore. Then, roosters crow from a farm. Thrushes and peewees, the earliest birds of morning, tune up.

He ties up the boat and taking his Mackinaw blanket over his shoulder, crawls into the tent. He checks his pocket watch. It is three fifteen. The others are asleep—or pretending.[43]

THE SURVEY WORKED ITS WAY around the northeast shore of the Lower Peninsula, stopping at Alpena, only three years old, which Newton thought "thrifty, intelligent and very promising,"[44] where they shipped a load of rock specimens home. The next stop was Mackinaw Island, where the party shipped more specimens and collected mail. Newton found he had mail "most interesting" waiting for him.[45]

Newton maintained connection with the many females in his life through these anticipated mail stops. His mother, sister Antoinette, Nettie Watkins of Grass Lake, Laura Bates from back East, Jennie Lines of Ann Arbor (Julia Winchell's sister)—all sent letters to the far-flung ports the survey visited. Newton replied to them all and mailed letters whenever he arrived at a port.

Newton found Mackinaw Island, the oldest settlement in the region, "the most enchanting place that we have met since we started out."[46] The town had had a Native American presence since 900 BCE and had been a key French fur-trading post. In 1860, it flourished as a tourist haven. Newton noted approvingly that a constant breeze kept mosquitoes at bay.

He was most intrigued by Mackinaw's multicultural atmosphere. He estimated that Native Americans comprised about a third of the population, and they were among the town's elite. They held government posts and contributed to civic life. White and Native American ladies with parasols mingled promenading the streets in fashionable gowns.[47]

After Mackinaw, the survey team explored the islands of Big and Little St. Martin, Bois Blanc, and Round, then moved on to Drummond Island to a quarry on its east end (the hard rock face not conducive to mosquitoes).[48] Leaving Drummond, they rowed against a headwind to the Canadian settlement of Bruce Mines, where Cornish immigrants had been mining copper since 1846.

Sundays were off-days for the crew. The men lounged in camp, reading or writing letters. On a Sunday spent in Sault Ste. Marie, Newton considered attending church, wanting to hear a preacher he knew from Ann Arbor, but decided not to. With his shaggy hair and sunburned skin, he felt unfit for society.[49]

Still, after a month and a half, life in the field suited Newton well. He had begun describing rocks as "Clinton [shale] of the Niagara group" or "conglomerate." He could recognize rock layers that corresponded to those of other locales. And he had eaten gull for supper and pronounced it good.

Newton, writing by candlelight each night in the tent, felt cozy and safe from the elements even in storms. His plant press—a good summer's work—formed the headboard to his bedroll. He kept books, paper, and pencil nearby, and plants to identify. At one camp he wrote, "my tent floor's complete tonight with *poison ivy* which is very common everywhere we have been this summer. It never affected me in the least although I never took trouble to avoid it."[50]

In late summer, the survey turned its attention to Lake Michigan's northern shore. At Little Traverse Bay, Newton was amazed by the great sand dunes that rose to more than a hundred feet high. He climbed the first dune he encountered and was stunned to see more dunes beyond it. Newton studied the dunes and surmised that they formed in the same way as snowdrifts. He thought that the sand was originally deposited as glacial till and then had been subjected to the effects of wind and water.[51] He was beginning to see with geological eyes.

The survey team encountered a Catholic missionary village, Vill a Cross (Cross Village, Michigan), a community of about two hundred, perched on a bluff overlooking Lake Michigan. The Protestant Newton was not impressed. "The inhabitants are principally Indians tyrannized by the priest," he observed.

There was also a nunnery three miles beyond the village inhabited by Native American women. Newton observed, "The free active daughters of the 'Forest Primeval' are taken from their rude, wild woods manner of living, to spend one of inactivity and seclusion, of possible [*sic*] tenfold more deplorable and despicable than their former [life.]"[52]

By the first week in September, the survey reached Grand Traverse Bay. There they collected few rocks but many plants. The Lake Michigan side of the Lower Peninsula had entirely different vegetation, dependent as it was on sandy soil. The spruces, cedars, and tamaracks of the eastern side were replaced by sugar maples, hemlock, and "monstrous oaks."

Newton returned with the crew to Mackinaw Island, and from Mackinaw they caught the *Cornet* to Detroit, and then the train carried them home to Ann Arbor. Newton would spend the next three months unpaid, keying out the remaining plants and writing up his report. He completed it the day after Christmas 1860 and turned it over to Alexander, his boss, that evening.[53]

THOUGH HE COULD SCARCELY AFFORD IT, Newton entered the University of Michigan for his sophomore year in October 1860. He still owed Uncle Lewis money and had nothing extra for the coming school year. He planned to attend college for a year, then teach school for a year, studying again on his own. Alternating work with schooling, he thought he could complete his degree in a reasonable time. He never earned enough, and he never had enough time to study while teaching. Nevertheless, he pushed doggedly on.

In spring 1861, Newton passed the year's final exams. He now possessed two years of a college education. He had been a standout student in analytical geometry and calculus. Prior to the exam, his professor assigned the class a problem that if solved successfully, exempted the student from the final. On examination day, Newton found that he alone had solved the problem.[54]

Although success in mathematics was a personal triumph, Newton pondered far weightier concerns that spring. On April 12, Confederate forces had fired on Fort Sumter, South Carolina, initiating the bloodiest four years in American history. By June, eleven states had seceded, forming the Confederacy. The country was at war. President Lincoln called for volunteers to help quell what was initially seen as a rebellion. Michigan was asked to provide one infantry regiment (one thousand men) that was fully armed, clothed, and otherwise equipped to aid the federal government. Michigan's young men responded enthusiastically and soon organized ten companies of one hundred men each.

As state officers in the vicinity of Ann Arbor recruited soldiers, Newton offered his services as a drillmaster. Since a schoolmaster is accustomed to drilling students, Newton was deemed competent and hired on the spot to train troops at Camp Fountain, northwest of town. Through bureaucratic miscommunication, he was not paid, though he agreed to continue his drills.[55]

Loans from family members piled up. Newton owed Uncle Lewis, and he owed his younger brother Robby. He learned he could enter the army as a lieutenant. First lieutenants earned $108 a month. But they had to pay for their uniform ($25), sword ($25), and revolver ($25.) There was no way Newton could scrounge up that kind of money. He did not receive a bit of encouragement from home. Even his two younger brothers wrote, urging him not to enter the army. Newton was undeterred, noting, "If I had always acted on their advice I should never have done anything, should never have left home."[56]

As he pondered how to raise the money, Newton continued to drill at Camp Fountain. He occasionally drew guard duty at camp. One fateful night, in a chilly rain, he dealt with four intoxicated recruits, hell-bent on destruction. "They were just drunk enough to be ugly and ready to fight any person who crossed their path," Newton recalled later. "They were bound to tear everything all to pieces."[57] Newton, wearing a (borrowed) sword, unsheathed it and brandished it boldly, thinking to intimidate the men. He handcuffed one and shackled another and tied him to a post to prevent him from attacking the prison guards. He then confined them until they sobered up. Total arrests that evening numbered seventeen.

Off duty at nine o'clock the next morning, he went home to sleep, but wound up from the previous night, he could not drift off. His throat was sore, and as the illness progressed, he suspected that he had caught diphtheria. The disease had ravaged Alexander's family earlier in the month, and it had killed his infant son.[58]

In addition, Newton contracted typhoid fever. His physician, the medical officer at Camp Fountain, had seen a number of cases. Left untreated—and there was no treatment in 1861—typhoid fever lasts four weeks, with four identifiable stages. In all, Newton was confined to bed for five weeks, feverish and delirious. In the crucial third week, the family set up "watchers" to sit through the night at the bedside. Alexander and Julia nursed Newton through the potentially fatal illnesses. His fraternity brothers in Delta Kappa Epsilon served as watchers, rotating through the nocturnal vigils.[59]

Typhoid fever leaves its victims emaciated. Newton also lost hair. But when he began to recover, he quickly bounced back. Back home, his parents convinced Uncle Lewis to loan Newton whatever funds he needed for college, so that military life did not entice him.[60] Newton, weakened and chastened, accepted the offer and enrolled at the university for his junior year. Gone were his military aspirations.

THROUGHOUT THE LONG MONTHS OF WORK for the geological survey, his sophomore year of college, flirtation with the army, and serious illness, Newton had maintained his engagement to Nettie Watkins of Grass Lake. She, too, suffered from typhoid fever, but letters went back and forth between them. At the end of his junior year, Newton made plans to teach—"am sadly in want of money to get a coat, pants, socks and boot repaired as well as a book or two," he confided to his journal.[61] Nettie enrolled at Albion College.[62]

In August, Newton landed a plum teaching job in Flint, Michigan, hired as principal in charge of three departments and 150 students. He also taught a mind-boggling array of subjects: geometry, astronomy, French, universal history, and Latin, as well as the core topics of geography, grammar, arithmetic, reading, writing, spelling, composition, rhetoric, and singing—led by the musical Newton, of course.

Flint in 1862 was less than half the size of Ann Arbor, but the city was located on the Saginaw Trail, a long-established route from Detroit to Saginaw. Deforestation of Michigan's rich white pine stands had benefited Flint, which flourished with new industry. Newton's new school was one of three in Flint.

The class that afforded Newton the most pleasure was French, comprised of twelve young women and three men. They met after the regular school day and were mature enough to thoroughly engage in their lessons and class discussion. Newton led them through French grammar and Farquelle's *Fables*.

Newton boarded that year with a Methodist minister, whose daughter taught the primary grades under Newton. She was a social activist, intellectually inclined, and above all, kind. Newton formed a strong bond with her, but she was engaged to an army physician, as he was to Nettie—"or I don't know what would have happened," he confided to his journal.[63]

When the school year ended in May 1863, Newton fully intended to return to Ann Arbor and complete his university degree. Then fatefully, he changed his mind and applied for a public school principal position at St. Clair, Michigan. This was a much bigger school than at Flint—four hundred students, with five departments and a salary of $700, more than Newton had ever earned.

The five departments were taught by women; Newton shared responsibilities at the high school with a woman, Charlotte Imus, an Albion graduate who had previously taught at the college. Newton had supervisory duties as well as a teaching load. He taught college-bound students, he was in charge of the school's five-hundred-volume library, and he founded a literary society for the benefit of his advanced students.

In addition, he taught an advanced class on Virgil after school, with some of the other teachers attending, including Charlotte Imus. Newton found her a fine scholar, able, cultivated, and principled. She was also dark haired, fine boned, petite, and pretty.

In May 1864, the tragic run of luck that haunted the Winchell family continued. Martin Winchell was gunned down by Confederate guerillas in Louisiana. The incident was unexpected and shocking. Martin had been residing at a Union-held Louisiana plantation to oversee planting done by African American hands, working at government-set wages. Troops guarded the plantation, but guerillas had lured them from their posts and stormed the house, killing the boss, Martin. A brother who had mentored Newton was gone.

At the end of the school year, Newton had saved enough money to repay most of his debts, with some left over to loan to both younger brothers to further their own educations. Robby was in Ann Arbor as a freshman at the university. Charley would follow him there in due time.[64] And in a clearing of the slate, he broke off his engagement with Nettie Watkins. The romance, for him, had run its course. They had quarreled, and he decided that there were dimensions of the relationship that he just couldn't live with.[65]

That was not the whole story. Newton had become smitten with the brainy Charlotte Imus, and she with him. Their courtship was discreet. They saw each other at school, after school, in church, and on picnics. They took a buggy to Lake Huron with a picnic lunch and a spyglass to train on boats out on the water. They skipped stones. They gathered chestnuts. They walked the beach.[66]

Later that month, they made plans to marry. He was twenty-four, Charlotte was twenty-seven.

METAMORPHOSIS

1864–70

NEWTON WINCHELL'S WORLD EXPLODED with joy the summer of 1864. He and Charlotte Imus—he called her "Lottie"—would marry before the beginning of the next school term. His happiness glowed in his letters. When he arrived back in Ann Arbor after the close of school, his brother greeted him with, "What? Did you come *alone*?" Fully expecting a visit from a future sister-in-law, Julia Winchell echoed her husband: "*Alone?*"[1] But Lottie had bypassed Ann Arbor for the time being—Alexander was leaving, and the Winchell household was busy with packing.

Letters between Ann Arbor and Lottie's home in Galesburg, Michigan, hastened back and forth in July. Newton tried to restrain himself but could not. He frequented the post office, waiting for mail to be sorted and slipped into slots. He calculated the time a letter would take in transit and how long it would take Lottie to reply. He snatched envelopes eagerly, tore them open, and devoured the letters as he walked home. Read them again later in the day, and a third and fourth time before bed. "Good night," he signed off to his intended one evening. "It may be . . . late before I get myself entirely gone from you so that I can sleep."[2]

At Lottie's suggestion, Newton read Elizabeth Barrett Browning that summer. The poet's masterpiece, *Aurora Leigh,* is a love story with a strong woman narrator. The epic poem ends with the two lovers acknowledging that their individual vocations are equally important and that they need each other to flourish. Newton told Lottie that he especially liked the ending, where the lovers are reunited in joy.[3]

Aurora Leigh was an indirect indication of what life with Lottie would be like, and he would in the ensuing years come to articulate what instinctively drew him to this intellectual woman. "I like a wife . . . [who] is capable of conversation and thought on the great topics that concern us, one also that maintains an interest in my own affairs, and can be a companion in every sense."[4]

In midmonth, the two arranged a visit to the Imus farm in Galesburg so Newton could meet her parents. He must have passed muster. At the height of the growing season, Alonzo Imus went so far as to lend the young couple his team of horses, Brownie and Foxy, so they could ramble the countryside alone together.

But Newton failed to find the right time to ask her father for his consent to marry Lottie. Understandably nervous, he postponed the conversation until the very last moment. Then he discovered to his dismay that Alonzo would not be accompanying him to the train depot to see him off.[5] A letter not asking so much as stating their intent to marry would have to do. Newton wrote to Lottie explaining this, closing with "Hope you are well. Hope you are all well. Hope Brownie and Foxy are well. Hope you will write to me soon. Hope I shall hear from you tomorrow. . . . Hope you believe and know me yours ever."[6] Another letter, to Alonzo, was enclosed.

Wedding bells rang from the Galesburg Methodist–Episcopal Church for Newton and Lottie three weeks later. On Wednesday, August 24, 1864, at eleven thirty in the morning, a former colleague of Charlotte at Albion College pronounced them husband and wife. The first stop on the couple's honeymoon that day was Albion, where Lottie's friends, former pupils, and fellow teachers showered them with gifts and well wishes. Newton thought the reception in a gas-lit parlor interminable. Finally, his bride led him away up the stairs and murmured, "First door on the right." The August night was sultry. Seven years later on their anniversary, he would remember that the boardinghouse bed squeaked.[7]

"MARRIED AND KEEPING HOUSE! Such a change has actually taken place," Newton recorded in his journal in September.[8] He was back at his post of superintendent of the St. Clair, Michigan, public schools. Lottie, having resigned her position the previous term, now tended to domestic matters. The newlyweds were not alone. Her younger sister Henriette, teaching in Lottie's stead, lived with them, as did her younger brother, Henry. Robertson Winchell, "Robby," also dropped in on occasion. Following his older brothers' example, he had completed his freshman year at the University of Michigan and was teaching in nearby Birmingham that school year to earn money.[9]

The second year of a teaching position is always easier than the first. With lesson plans sketched out and a wife to oversee housekeeping matters, Newton found time to study toward completion of his baccalaureate degree. He had one year remaining. He reopened his German textbook, brushed off his knowledge of Greek, and made headway in his mastery of geology. Newton kept a little volume filled with quotes he wanted to remember, and James Dana's *Manual of Geology* was the source of many tidbits[10]—on glaciers, climate change, duration of the various geological ages, and the diversification of fossils.[11] He hoped to gain enough competence on his own so that he could pass his college exams without additional coursework and receive his diploma in the following year, 1866.[12]

Lottie became pregnant, and the impending birth complicated matters. Newton felt the press of responsibility. He had hoped to instill a little adventure into his life by finding work in the South. Instead, he looked for a suitable local job. He landed one at Kalamazoo, the first of four successive teaching posts of short duration. Kalamazoo had the benefit of being only nine miles from Galesburg, linked by a railroad. Lottie could stay with her mother in her last weeks of pregnancy.

Hot weather lingered when the school year began in early September. Newton had rented a tiny room, eight feet by eight feet, in a boardinghouse in downtown Kalamazoo. The first night, he trudged home—no Lottie waiting, no letter in the postbox—peeled off boots first, then pants and socks, took up pencil and paper, and lay down on the floor to write. The miniscule digs barely had room for the narrow bed, his trunk, and a washstand. Later, he would kneel at his bed, placing the kerosene lamp on the covers to write, until discovered by his alarmed landlady, who then somehow wedged a little table into the cramped space. He could no longer fully open the door, but it was less likely he would set the place on fire.[13]

The pay was poor in Kalamazoo, and the job description broad. Newton taught all the classical subjects: algebra, geometry, Latin,

Greek, readings of Cicero and Caesar, and astronomy; he added natural philosophy, focusing on Newtonian physics, as well as a class in chemistry and one in geology. There was a small collection of rocks and fossils in the school, and Newton added his own specimens so his students got hands-on experience.[14] The high demands of the job coupled with meager pay, less than he had been led to believe, fed his growing dissatisfaction with teaching. The frugal Winchells could barely live on eighteen dollars a week. They accumulated debt, and Newton still had school loans outstanding.[15]

Newton and Lottie's first baby was born on November 1, 1865. Lottie gave birth on the farm, and Newton took the train from school to be with her. In the weeks before the baby came, Newton, lonely in Kalamazoo, dreamed of a child with little black "twinklers" and dark hair.[16] It was a hopeful sign, but as the due date approached, he was filled with dread about mishaps. Mindful of his sister Laura, who had died in childbirth with her first baby, he termed the incipient labor and delivery "a crisis" that could only be entrusted to God's care.[17]

Lottie was adamant in having a homeopathic practitioner attend her—one who, Newton in retrospect decided, "proved to be unpracticed and ignorant."[18] A second, experienced homeopath from Kalamazoo was finally called, and the baby survived, despite the prolonged labor and delivery and having cold water thrown on him when he stopped breathing after birth. (The second practitioner did mouth-to-mouth resuscitation to revive the boy.)[19] He weighed nine and a half pounds, and they called him Horace after his paternal grandfather, "Hortie" for short.

Lottie and Hortie moved to a boardinghouse in Kalamazoo to be with Newton—his salary could not afford a house. In the following months, the young parents were continually amazed by the brilliance of their child: He cuts a tooth! He sits up in his crib! Shakes a rattle! Newton kept a written record of developmental milestones, weight gain, illnesses, and the medications the family used.[20]

As summer grew into warmth and greenness, Newton finished his duties in Kalamazoo and traveled to Ann Arbor to consult with Alexander. Through letters, Alexander had offered him a position as his assistant, in either Ann Arbor or at Lexington, Kentucky, where Alexander had been offered an additional professorship of natural science at the university. (He also intended to keep his Michigan professorship.)

The two brothers were ships passing in the night—Alexander was in Kentucky when Newton got to Ann Arbor, and Newton left before his brother returned. The job fell through. However, Newton's time in Ann Arbor was not ill spent. He was able to take final exams and complete his baccalaureate degree, eight years after entering college as a freshman.[21] It had been a dogged, persistent slog.

Lacking a job, Newton once more scrambled for employment. He found a position in Colon, Michigan, a tiny town not on a rail line, which exacerbated its isolation and provincialism. Newton termed it "a vicious little village . . . filled with little-souled men, with a few noble exceptions that prevent them from reverting to barbarism."[22] The pay was terrible, but the cost of living low. Newton and Lottie made a few friends, and the family eked by. Newton served not only as teacher but as principal and janitor to the school. He left Colon in April, before the end of spring term—he had to get out of that soul-sapping job—and took a substitute position in Port Huron to complete the school year.

Adversity sometimes clarifies the mind, and it did in this case. By May 1867, he had a new understanding of what he wanted to do in life. He wanted to work as Alexander did, in natural history.

With Lottie and Hortie still in Colon, Newton spent several weeks in Ann Arbor in summer 1867, working with Alexander in his laboratory. He learned some protocol for handling specimens, and various tests to determine chemical composition of rocks. He fully intended to move to Lexington, Kentucky, that fall to finally serve as Alexander's assistant at the university, when an unexpected job offer came through. Adrian, Michigan, in the far southeastern corner of the state was interested in him for the post of superintendent of schools. The pay was high, and the community an attractive one. Newton turned them down flat: his future was in Kentucky. Later that night, though, he reconsidered. How wonderful it would be to have enough money for once and how much easier for his family to remain in Michigan. He notified the school board he had changed his mind.

In doing so, Newton delayed his career plans for the good of his family, which now numbered two children. While he had been away, Lottie had given birth to a healthy baby girl on May 22, 1867. They called her Ima Caroline, after Lottie's maiden name and Newton's mother.

Unlike Hortie, who resembled his mother, Ima looked like him—everybody said so. She was "bright, good-natured, blue-eyed and fair-skinned,"[23] an unmistakable Winchell. But, he noted in his journal, she had her mother's feminine hands and feet. Once again, Newton had dreamed of a child, a girl with blue eyes and blond hair, shortly before the birth.[24]

Before school began, in the heat of August, Newton moved the household—two babies, a convalescing wife, and all their earthly goods—to Adrian, where they bought furniture—for the first time—and moved into their rented house. Newton and Lottie loved Adrian. Their nearest neighbors, Dr. and Mrs. Stephenson, became close friends, Newton oversaw the expansion of the school system, including a new building, and the community sent off graduates to the university. Younger brother Charley came to Adrian to take the college prep courses at the high school.[25] Lottie regained her good health, and though thoroughly occupied by two young children, she organized a "Ladies Library," with holdings of a thousand volumes and $1,500 in the bank. All the prominent families in town subscribed to it, and Newton thought the library would be a lasting legacy of their time in Adrian.[26] He noted in his journal that the project had the active opposition of the Young Men's Christian Association, adding darkly that the organization was Christian in name only.[27]

AFTER TWO YEARS IN ADRIAN, Newton left the superintendent's job and became the assistant state geologist of Michigan in Ann Arbor. His brother Alexander was his boss. The Michigan state legislature had defunded the state geological survey at the start of the Civil War but had reinstated the money in 1869 in time for the summer field season.

Survey work had its drawbacks. The state survey was entirely dependent upon funding from the legislature, which gave the work a political tinge and meant that money could be capriciously withdrawn without regard to the status of the work.

Newton had wearied of public education, though, and after the Adrian job, he would never work again in the public schools. Born to teaching and for years never deeply questioning whether to continue in it as a career, he became worn down by the miserly public attitude toward teachers, the stingy salaries, the bleeding of the school days into evening hours. There were always papers to read, exercises to correct, and lessons to prepare. Even the higher-paying Adrian position was not enough for his family to live well and to pay off the long-standing college loan to his uncle Lewis Winchell. Lewis had died in the interval, and Newton now owed his estate.

His long stint in public education left him with several gifts: broad knowledge in disciplines as diverse as French and mechanical physics, honed teaching skills, and vast experience in administration of budgets and paid staff, which would prove invaluable in future work.

In Ann Arbor, as a working geologist, Newton finally made good on his school loans, and even though the salary was lower than that in Adrian, he was able to write in his journal, "I . . . have some time to study and work for myself. I enjoy this much better than I did superintending school."[28] The study time and his work under Alexander allowed him to complete the requirements for a master's degree in 1869.[29] Still, the lower salary meant that the Winchells needed to live a quiet life. They entertained very little and had no paid pew in the Methodist-Episcopal church.[30] Lottie did without hired help, and the family had not needed new clothes, so they were spared that expense. Also, following the precedent set by his older brothers, Newton now lent money to Charley, his youngest brother, who was a college student.

In spring 1870, Newton and Lottie took on younger brother Robby as a paying boarder, and Lottie used his rent to hire a girl to help. Hortie and Ima were four and two, and Newton would be leaving in late May for fieldwork for the survey with plans to be away most of the summer. Robby would join him in July as his assistant, just as Newton had assisted Alexander a decade earlier.

Fully immersed in family life, Newton thought constantly about his small children while away from them. He was accustomed to rocking the babies to sleep and singing to them, both nestling on his shoulders. His letters were filled with admonitions to Lottie to be watchful of their diet, and to Hortie to not hurt his baby sister, and with kisses by mail to Ima. He wrote poetry to ease the sadness of being apart.[31] But separation was part of the job. He would be in the field in summertime if he continued as a field geologist.

WORKING GEOLOGIST

1870–72

On the western shore of the Door Peninsula of Wisconsin, jutting into Lake Michigan, Newton Winchell pulls his chisel-tip rock hammer from his collecting bag. Standing up in the rocking boat, he gives the limestone outcrop rising from the water a carefully considered whack. On this twenty-first of June 1870, the summer solstice, he is revisiting a site at Little Sturgeon Bay that he and his crew stopped at four days before. He wants to collect additional rock specimens and have a second look at the lower layers of the Niagara Limestone that at intervals lines the shoreline of the bay. Niagara Limestone is a drab gray color; it yields easily to the hammer.

He has been in the field for over three weeks. The solstice sun has reddened and weathered his fair skin, though like all the crew, he never works without a hat. His dark beard is woolly, his hair needs a trim, but he's not overly concerned with appearances, for the survey seldom ventures into even the most rudimentary society in the small communities along the western shore of Lake Michigan.

Above, the verdant summer forest shimmers in the breeze. The clear water of Lake Michigan laps at the rocky shore. Through aqueous shades of teal and emerald, broken-off pieces of limestone can be seen on the lake bottom.

Winchell feels fortunate to have this hammer in his hand. Two days before, when he and the captain were underway in restless water, some hunch told him to stow it in his collecting bag and buckle the bag to a seat in the Mackinaw boat. Perhaps it was his growing lack of confidence in the captain, whom he had hired. Perhaps the horizon seemed a shade too dark for early afternoon. By two o'clock that day, a menacing thunderstorm loomed. They were in this very bay when the squall struck, and within minutes they capsized under a terrific wind. Newton and the captain successfully righted the boat and steered it, with one remaining oar, to a sand beach on the peninsula, but he had lost his sturdy field coat and everything that had been in its pockets—most significantly, his clinometer compass worth eight dollars. Fortunately, he had shipped home several boxes of specimens immediately before the misadventure. His men spent yesterday combing the beach for articles that might have washed up after the spill, but found nothing.[1]

Winchell ponders the limestone and labels the specimens he is collecting. This limestone has a new layer, different than what he has seen at other sites this summer. Sketching in his field journal, he designates the lowest layer, a soft shale, as "A," the middle layer as "B," and the gray Niagara Limestone as "C." He has been following rock outcroppings of this type across the Upper Peninsula of Michigan. He mapped similar limestone cliffs on a narrow finger of land poking into Lake Michigan east of Escanaba, Michigan. The finger of land with its cliffs extended southward, lining a bay, and then seemed to disappear into Lake Michigan. Due south, along the Door Peninsula's western side, the limestone appears again and seems to line up, suggesting the two outcroppings are part of the same ancient ridge.

To get to this point, he and his crew sailed south from Escanaba along the Wisconsin side of Lake Michigan, into Green Bay, but did not see any outcroppings in Green Bay. However, here at Little Sturgeon Bay, it is exactly like that seen on that narrow finger of land, seventy-five miles north. On a map, it looks like it might once have been connected. Now, the rock types also suggest it.

Winchell scratches his head, piecing it together. There's no one to discuss rocks with in the collecting party. He has with him the captain—the one who capsized the boat Sunday—a cook, and an assistant, who is along on his own expense. He helps Winchell in camp, and he assists in collecting and caring for specimens, but he doesn't have geological training. In July, this assistant will leave when Robertson Winchell joins the party. Robby has no formal geological training, but he is a Winchell, with a fine Winchell mind.

AT THE START OF THE 1870 FIELD SEASON, when he was thirty, Newton took the train to Chicago to have a firsthand look at a limestone quarry southwest of the city. It had been pronounced "Niagara Limestone" after careful study by qualified geologists, but when Newton examined it, he decided that is was different from limestone he had seen in the past on the Michigan Survey.[2] His summer task was to complete the survey of "the South Michigan shore, including the region of Green Bay, and the Beavers, Fox and Manitou Islands"[3] Quarries like this one, and rock outcroppings that emerged from lakeshores or streambeds gave him an opportunity to actually see bedrock, which is normally hidden beneath the surface of the earth. Studying the exposed rock, the fossils it contained, the color, the texture, and the hardness made plain what kinds of rocks lay hidden. He would be able to hook these brief points of revelation together to construct a map of the rocks underlying Michigan.

Since he was paid by the State of Michigan, it would seem that land that was uncontestably part of Wisconsin should have been of no interest to him—and insofar as its economic value was concerned, it wasn't. But Newton Winchell was his own agent that summer. Alexander had given him no instruction and told him only to use his own judgment in how he conducted his part of the survey.[4] And Newton was after broadscale information. He was trying to define an ancient marine basin and the prominent rock ridge that encircled it.

The Niagara Escarpment—that prominent rock ridge—is one of the striking geological features of the Great Lakes region. The ridge starts in midcontinent, between Lakes Ontario and Erie, where it forms a ledge. Water that spills over the ledge is famous: Niagara Falls. The ridge can be traced to the west, arcing along the northern shore of Lake Huron and continuing along the northern shore of Lake Michigan. It encloses the Michigan Basin, what was once a large, shallow sea in the Paleozoic Era, 500 to 400 million years ago.

The rock layers of the escarpment were once a seabed. There are sandstones, which are compressed sand layers; shales, compressed clay; limestone; and dolomite. The last two are composed of carbonate containing the shells of animals that once lived in the ancient sea. Some of the animals, like trilobites, are long extinct. Others have descendants that still inhabit tropical seas, like corals, sea lilies (crinoids), snails (gastropods) with spiraled conical shells, and brachiopods, clam-like invertebrates with hinged shells.

Niagara Falls was first described scientifically in 1842, twenty-eight years before Winchell directed his party to Door County. He could build on that knowledge. Since then, geologists had been adding pieces to the jigsaw puzzle of the underlying geology of the continent. There was much to be learned of the continent's geological past.

Winchell was undertaking his research using several geological theories that we now take for granted but were not commonly held in 1870. One was that the earth was much older than the 6,000-plus years that literalists had calculated based on the generations outlined in the Bible. Scottish geologist James Hutton, considered the father of modern geology, had laid out an understanding of a very old, restless earth in 1788 in his seminal book *Theory of the Earth.* In it he pronounced the oft-repeated axiom: "[in geology] we find no vestige of a beginning,—no prospect of an end."[5] Newton's own father, Horace, who significantly influenced Newton's early education, did not believe this. Horace established the earth's age via biblical generations in his writings,[6] so Newton's leap in understanding must have been made at Ann Arbor, when he came under the intellectual influence of his brother Alexander. Nineteenth-century geologists varied in their estimates of the age of the earth. Charles Lyell thought it might be several hundred million years old; Lord Kelvin, less than a hundred million.[7] Though orders of magnitude older than the biblical estimate, these still fell far short of what we consider the age of the earth to be today: 4.6 billion years old.

Another tenet was that rock layers can be categorized by the types of fossils they contained and that in undisturbed beds, the oldest rocks were on the lowest level. This seems so obvious, but the law of superposition (oldest rocks on bottom) was proposed by the Danish scientist Nickolas Steno only in 1669. Typing rock beds by the fossils they contained wasn't done systematically until 1815, by Englishman

William Smith, which led to a geological map of Great Britain.[8] It was the first geological map to be constructed in this manner anywhere in the world. Winchell would use small invertebrate fossils in describing the sedimentary limestones of southern Minnesota.

The rocks of the Niagara Escarpment were formed in the Silurian Period, a time period fairly well studied in Great Britain. The textbook that Winchell consulted, James Dwight Dana's *Manual of Geology,*[9] notes that the subdivisions of the Silurian Period in North America and Europe are very different. Dana did not assign any age to the rock layers he described in his book, but he was very clear that the Silurian was a time of shallow, continental seas, indicating that the earth looked dramatically different long ago than it did in 1870. The idea of a dynamic, changing earth was a much different understanding of the world than Winchell had undoubtedly had as a boy.

Winchell knew that the rock layers he was observing were originally sediments from the sea and that they had been deposited horizontally, so a cutaway view would resemble a layer cake. This, too, was defined by Lyell's *Principles of Geology.* If the layers were tilted (dipping), there must be a reason for it, such as an event like mountain building or an earthquake. Winchell measured dips at each location at which he encountered limestone outcroppings, and with data accumulated was able to map the basin cradled by the rock.

Lastly, Winchell knew that an ice sheet had covered the land he was studying. The first geologist to propose an ice age was a European, Louis Agassiz of Switzerland, who published *Études sur les glaciers (Studies on Glaciers)* in 1840 in French. He accepted a post at Harvard in 1847 and continued his work on glaciers in North America.

As Winchell and his party labored in June 1870, they could see evidence of glaciation everywhere. Grooved rocks at the head of Green Bay and near Big Bay de Noc bore witness to a heavy sheet of ice pressing down rocks carried by the glacier, thus scratching bedrock. Indeed, the Green Bay valley seemed excavated by a glacier moving in the direction of the scratches, northeast to southwest. Because of the orientation of other rock grooves, or striae, Winchell surmised that the main glacier had originally moved in a north–south direction but had been deflected to a more northeast–southwest orientation when it encountered the Niagara Escarpment. Lastly, he thought that as the glacier retreated, it established an outlet draining Lake Superior into

Lake Michigan through a valley in the Upper Peninsula. This would mean, he noted, that the present outlet of Lake Superior at Sault Ste. Marie was fairly recent in origin.[10]

Laid out before his eyes were the effects of a great and mighty force, an enormous amount of ice that had changed the surface of the earth, its land and waters. With his hammer and compass, his careful field notes, and organized and meticulously labeled collections, Winchell was piecing together the restless history of the land.

WINCHELL HAD A TASTE OF HOW SUDDENLY fortunes can turn for a survey geologist when the Michigan legislature unexpectedly terminated Alexander Winchell as survey head and cut off funding. Newton, as the assistant, was also out of a job.[11] Once again, the father of two, with a now-pregnant wife, was left to scramble for money. Earning cash any way he could, Newton sold some magazine articles on building homes in Michigan, and others on agricultural prospects in the northern Lower Peninsula.[12]

Scrambling some more, he signed on to the Ohio Geological Survey for the coming field season in summer 1871. The highly respected John S. Newberry headed the survey.[13] Ohio was considerably more settled than its neighbor to the northwest, and this was much different work than that in Michigan. Rather than tromping with hired men and guides through bogs and forests, or sailing on massive waters in a small boat, he took a train to the nearest town in the county he was assigned to survey, put up at a local hotel, and walked or rented a horse and buggy to reach rock outcroppings and quarries. He worked alone, taking measurements and filling several volumes of field journals with neat, precise descriptions and sketches of what he was observing. His boss, based in Cleveland, assigned him twelve counties in the western part of the state, which Newton systematically described. Lottie and the children remained at home in Ann Arbor, with hired help to aid her. Alexander's family was only blocks away.

That spring of 1871, Alexander, who also needed employment, took a short-term job in Minnesota. The state legislature wanted a geologist to assess possible salt deposits in Belle Plaine, along the Minnesota River in the southern part of the state. Alexander concluded that the prospects for commercial salt production at the site were dim,[14] but he returned to Ann Arbor with brighter news for Newton: he had caught wind that Minnesota might have a prospective job for him.[15]

The following winter at home in Ann Arbor, though no longer an employee of the State of Michigan nor a student at the university, Newton somehow had access to the school's chemistry lab. He embarked on a study of chemistry, did analytical work in the lab, and attended lectures sporadically.[16] He also used the time to write up his work for the Ohio Survey. The reports were delivered on time, and he was rehired for the coming 1872 season.

Despite setbacks, Winchell continued to find employment in the field. As development of the American West ripped across the landscape, opportunities for geological work bloomed. Early in spring 1872, he was hired to examine a copper mine near Silver City, New Mexico. Newton had seen copper mining in his first summer with the Michigan Geological Survey at Bruce Mines, Ontario, one of the first commercial copper mines in North America. The New Mexico work was of short duration and paid well, $380 for thirty-eight days of work. A trip into the unsettled desert Southwest gave Winchell the chance to see firsthand an area he had only read about.

Lottie had given birth to a second daughter, Avis, on November 1, 1871. Newton now left his competent wife to direct domestic affairs with three small children, ages six, five, and seven months. Life without Newton for several long months was becoming routine for Lottie.

Winchell traveled in a party of four, first by rail through Missouri and then by stagecoach beyond Kit Carson City, Colorado. He had not been west before and marveled at the black prairie, its treeless aspect, and in Kansas its lack of surface features, which he termed "monotonous."[17] Bones from slaughtered bison that had been left to rot lay scattered across the countryside.

Around Cimarron, New Mexico, the party encountered over a hundred Jicarilla Apache Indians entering town to receive government rations.[18] In a letter to Lottie, he described their "scrawny Mexican ponies," their clothing, each person with a blanket, and weapons (chiefly bows and arrows). He observed that the women carried more blankets or buffalo robes than the men, in which they wrapped and strapped their children. Older children were set two or three on ponies; mothers, sitting astride horses, fastened small children to their backs. Men, women, and children painted their faces in red, green, yellow, white, and black paint. He noted their jewelry—one young man had at least ten bracelets—and their great fascination with handheld mirrors, which he thought they used as he did: "in fixing and painting."[19]

Winchell's adventures in the West became more exciting when the stagecoach reached Fort Craig, south of Santa Fe. There the party heard persistent rumors of Apache hostilities. A general anxiety pervaded the whites at the fort. Newton didn't have a gun, and he wanted one.[20] He wrote Lottie that the Apache had guns and that they "lie in ambush," also that a slow-moving stagecoach was at a "great disadvantage." Underscoring his worry, he signed his letter "Good bye. Yours ever," rather than the standard "Yours affectionately."[21] One could imagine Lottie tossing and turning in bed the night she received that letter. It had been only eight years since Newton's brother Martin had been killed by ambush in Louisiana.

A mere thirty-six hours later, a sheepish Newton wrote that he might have overreacted. Life was calm in Las Cruces, and no one expected any trouble. The cacti were blooming, and he was in good health and spirits.[22]

Winchell finished up the New Mexico assessment within the week and made arrangements to head home. He wanted to travel through Denver, hoping to see the Rocky Mountains and the legendary Colorado River near Pike's Peak.[23]

Winchell arrived in Ann Arbor at the end of May and immediately headed for Ohio to resume his work for the survey. While he had been in New Mexico, the president of the University of Minnesota, William Watts Folwell, had quietly been collecting letters recommending the young geologist, before offering him a job to lead Minnesota's geological survey. All, especially Newberry's,[24] were complimentary, although one writer noted that Winchell was perhaps "hypercritical in his distinctions among the rocks . . . [although this is] not a demerit," he hastily noted.[25]

As June concluded, Winchell was methodically checking off the western counties in Ohio that he had finished surveying and thinking through an original theory of glacial till. He knew that he was in contention for the Minnesota job but thought that he was at a disadvantage, since he had not met any of the board of regents, only President Folwell.[26] He was thinking ahead to plan B—working under Newberry in Ohio in the winter—and plan C—working under his brother in Ann Arbor, studying chemistry—although by his own admission, he found the elder Winchell condescending and dismissive of his ideas.[27]

Newton was beginning to distinguish himself from his brother intellectually in other ways. When he learned that Alexander was having trouble with his college students, Newton observed that his brother was too theoretical and metaphysical. Alexander could benefit from more physical fieldwork—perhaps with him, in Ohio—or a diversion, like travel in the West.[28]

Folwell's letter came after the Fourth of July. Could Winchell travel to Minnesota and meet with the university's board of regents?

Within the week, Newton Winchell found himself at the Nicollet House in downtown Minneapolis, the best hotel in the young city. He walked across the Mississippi River to the neighborhood adjacent to the university and visited Folwell in his home. The job was his, Folwell told him; the regents had just wanted to see him, that was all. Could he start immediately? Folwell thought the survey would take twenty years.[29]

SETTLING IN MINNEAPOLIS

1872–73

July 13, 1872

Newton Winchell strolls through the neighborhood of Minneapolis–East District. He has crossed over the Mississippi River on a slender wooden bridge suspended by two cables, and flanked at both ends by two towers. The bridge, which conveys traffic between both portions of Minneapolis, hangs close enough to the water to be thrilling. He notes with interest the banks of the river, its swift current, the wooded islands directly above St. Anthony Falls. The tumult of the cataract and the bustle of the mills dominate the business district.

But here in the neighborhood, the East District is more serene. He and Lottie had thought that should Newton get the job at Minnesota, they would be moving to a raw, primitive town, a step down from gentrified Ann Arbor. Newton is pleased to see with his own eyes that this is not the case. Minneapolis is much larger than Ann Arbor and does two or three times the business. It has first-class stores and hotels. He's staying at the Nicollet House, with Belgian carpets, lace curtains, and an observation cupola.

He heads toward the university, which borders the neighborhood to the south. Classes are not in session in July, but workmen labor on a new wing to the one building on campus. Walking back toward town, Newton sees that new construction is occurring everywhere. Many of the houses are new, their lawns shaded by bur oaks that are remnants of the original oak savannah. None are old or dilapidated, and all are freshly painted. Streets being laid out extend into prairie grasses. Everything is very picturesque.

MAPLES ALONG THE MISSISSIPPI RIVER GORGE flamed in burnished hues, and Minneapolis residents anticipated autumn's onset as Newton and Lottie Winchell arrived in the city in September 1872. The state fair was over, and a presidential campaign under way, pitting newspaperman Horace Greeley against the incumbent, General Ulysses S. Grant. The summer had been one of "unbounded prosperity" according to the *Minneapolis Tribune*. Nothing could stop the city's advancement "on the bright highway to metropolitan greatness."[1]

P. T. Barnum's circus was in town with six giant tents, a live giraffe, and a horse-riding goat, Alexis. Proclaiming the circus "The Greatest Show on Earth," Barnum's ads asserted that "nothing like it [has] ever [been] seen since the world began.[2] Had he been asked, Newton Winchell might well have begged to differ. The work he would produce, the work that brought him to Minnesota, would in the years to come tell a greater story, that of the immense forces of nature, of rocks and fire and ice and time—unimaginable expanses of time—that few in the young state had pondered.

The Winchells moved into a rental house at 1223 Fifth Street Southeast, near the University of Minnesota. Winchell had procured it when he interviewed for the job in July. When the household goods arrived by rail, university president William Watts Folwell lent a maid to help the family settle in.

Newton and Lottie's new home was across Tuttle Creek from campus, a convenient walk. Other professors were neighbors; President Folwell lived seven blocks toward town. A boardinghouse for male students stood on Fourth Street between Thirteenth and Fourteenth Avenues, built and maintained by the head of the university's board of regents, John S. Pillsbury. A new public school on Fourth Street, only three blocks away, had recently opened. It would be an easy walk for Hortie, who would turn eight in November.

The East District of Minneapolis, formerly known as St. Anthony, a city of five thousand, had seamlessly merged with larger Minneapolis, on the west side of the river, only six months before. The merger put the combined city at nearly twenty thousand people, rivaling the capital, St. Paul, for the state's largest metropolis.

Businesses in the East District had grown up around St. Anthony Falls, the only significant waterfall on the Mississippi River. By 1872, entrepreneurs had harnessed the waterpower to drive a bevy of mills clustered near the booming cascade. Some businesses dug canals parallel to the river to capture its force. The waterpower seemed limitless.

But it was not. In 1869, a final attempt to expand capacity threatened the integrity of the waterfall's ledge. The entire milling district had been jeopardized. An apron over the lip provided a temporary fix. A permanent solution would be found a decade later with the intervention of the Army Corps of Engineers, and federal, state, and city tax dollars.[3] He did not know it yet, but Winchell would be called upon for his expertise.

Another disaster hit St. Anthony's milling district in 1870. A whole row of sawmills lining the river ignited in a blaze that could be seen ten miles away in St. Paul.[4] Sites along the river, however, were still prime property. They quickly sold and were redeveloped, though not as sawmills. The waterfront was a busy place, with factories and small mills. The dominant, now-iconic Pillsbury mill, soon to rise to prominence on the river's east side, however, was in its infancy.[5] The Pillsbury presence was felt in a different way: John S. Pillsbury owned a large, prosperous hardware store on Main Street, the site of the future Pillsbury A Mill.[6]

Minneapolis–East District was a self-sufficient community in 1872, with four schools, three fire stations, a post office, a bank, and several hotels, although the five-story limestone matriarch, the Winslow House, had closed in 1861. It reopened in 1872 as a joint home to Minnesota College Hospital, a forerunner of the university's medical school, and Macalester College.[7] Many a photograph of the expanding city was taken from the Winslow House cupola. A Universalist church anchored the district up the riverbank from Main Street. French Catholics would buy the building and add a steeple in 1877.[8] But in 1872, it was an unadorned, neoclassic building, crafted from the bluish Platteville limestone that was quarried at Spirit Island, one of five small islands immediately upstream from the falls.

First Congregational Church occupied a modest frame building on East Hennepin Avenue and Fourth Street. A Methodist–Episcopal church, also frame, stood a few blocks away on University Avenue at Central. The Winchells would join the congregation in January 1873 and figure prominently in its activities for fifty years.

When Newton Winchell arrived on campus that fall, the university was already in session. He was not on the roster to teach classes that term. Rather, he wore his state geologist hat. The Minnesota legislature had stipulated that the geological survey be conducted under the auspices of the University of Minnesota. From the beginning, Winchell was given rooms at the university for storage of specimens, laboratory space, and (theoretically) space for a natural history museum, though in 1872, Old Main was overutilized. Everyone was waiting for the new wing to be completed.

A scientific geological survey of the state had been proposed and vigorously supported by the university's president, Folwell. A thoughtful scholar, Folwell was thirty-nine years old in 1872, only seven years older than Winchell. He oversaw the legislation in March 1872 that established the survey and lobbied for oversight to be given to his school, the state's foremost academic institution. In this, Minnesota's geological survey differed from other state surveys, which were overseen by legislatures. Legislators granting funding were primarily interested in learning what resources (ores, building materials, agricultural, fossil fuel) might await development.[9] The setup of the Minnesota survey meant that it would be weighted more toward scientific discovery and less subject to the whims of politicians. It was also true that in the past, the governor had made several bad choices in appointing survey geologists and may have been only too willing to hand off the job to Folwell.[10]

The enabling legislation minutely described the purpose of the survey. The legislators wanted "a complete account of the mineral kingdom as represented in the state, including the number, order, dip, and magnitude of the several geological strata, their richness in ores, corals, clays, peats, salines, and mineral waters, marls, cements, building stones and other useful materials, the value of said substances for economical purposes and their accessibility."[11] Lawmakers also wanted their state geologist to run chemical analyses on the various rocks, soils, and ores and keep the results on file for future reference. This level of detail was also unusual for a state survey.

The legislature also called for a survey of the plant and animal kingdoms, with equal specificity. In addition, the director of the survey was to compile information on the weather, especially variations in temperature and rainfall. Since farming was a keystone of the state's economy, such information was essential.

Lastly, the legislation established a natural history museum, to be housed at the university. The collections of geology, botany, and zoology would be prepared and properly displayed—not only as teaching material in university courses but for the edification of all Minnesotans. The museum was to be heated, furnished, and accessible to the public.

Winchell faced a big job description, but at thirty-two, he possessed the tenacity, boundless energy, and intellectual vigor that already marked his work as a geologist. Within a week of arrival, he bought an account book, two pocket field books, and a large wall map of Minnesota. He then sent letters to all twelve railroad presidents with lines operating in the state,[12] requesting passes so he could travel throughout. He also ordered stationery with a new letterhead at the *Minneapolis Tribune,* which printed individual orders besides the newspaper, and then, with one hundred dollars in cash from John Pillsbury and the board of regents, he began.

As he had on the Michigan and Ohio state surveys, Winchell used information from local residents as a starting point. In early October, he caught a train for Winona, where immense limestone bluffs rise above the Mississippi River. His contact in town was Professor William Phelps, the head of the normal school there.

Normal schools, or teacher-training schools, produced the schoolteachers that labored in the one-room schoolhouses dotting the frontier. Winona's was established in 1860, the first tax-funded school west of the Mississippi River. The college occupied an elegant five-story building with a soaring clock tower that dominated the red-brick river town. Phelps, who had been with the school since its inception, had an extensive collection of minerals. His collection and writings established him as a close observer, and his geological work helped Winchell in drawing up a preliminary map that fall.[13] Phelps donated to Winchell's nascent natural history museum some trilobite casts, which Winchell deemed among the most perfect ever found in the state.

From Winona, Winchell boarded a train to St. Charles, and from St. Charles to Dodge Center, and then home to Minneapolis, geologizing at every stop. At St. Charles he met another valuable source of geological information, William Hurlbut, a former mayor of Rochester. Winchell called Hurlbut a pioneer of scientific development in the geology of southern Minnesota and was gratified by his able assistance.

It was a weeklong, get-acquainted tour, the first of five trips he made that fall, as he saw the varied rock formations, made contacts, and built relationships. On other trips he visited Stillwater and followed the St. Croix River north to the basalt outcroppings at the Dalles at Taylors Falls, then took a stagecoach to Stacy and caught the train home. He traveled to St. Peter and New Ulm, stopping at quarries in Kasota and at St. Peter's "Asylum farm," the forerunner of the state psychiatric hospital.

As he traveled, the young geologist noted prominent buildings and opined on their construction. The Stillwater courthouse, he observed, was of locally made red brick and trimmed in limestone quarried nearby. But it, and several Stillwater churches, also used bluish-colored limestone from St. Paul, which was "very attractive" but "a great mistake," he thought, since it would weather more quickly and chip off.[14]

A permanent snow on November 12 ended Winchell's first field season. He retreated to his desk and laboratory to read the geological reports that had preceded his, and to examine the specimens he had collected on his jaunts. His first annual report to his bosses, the board of regents, included a summary of the findings of previous geological work in Minnesota, an account of his initial assessment of the state's bedrock, plus a preliminary geological map.

This report wonderfully displays how Winchell's mind approached his pioneering work. With a schoolteacher's inclination to instruct, he informed the board of the advances in geology in the past half century. There were two guiding principles at work in the discipline: first, that the natural laws, as scientists understood them in 1872, also applied in the past; and second, that the past, the age of the earth, is immense. "Time is long," he observed. With these two postulates, "the geologist . . . may read in the rocks the grand changes the earth has undergone since 'the beginning.'"[15] Winchell added a chart depicting geologic time and the names of the rock layers in North America, Europe, and three midwestern states, Illinois, Iowa, and Minnesota. North America's middle was geologically unknown. Winchell intended in this chart to show the relationship of Minnesota to the "great Geological Series of the Earth," linking Minnesota, for the first time, to the rest of the world.[16]

Winchell had recommendations for the board of regents. He deemed the salary of $1,000 a year too low to adequately carry out the survey. He suggested that certain lands thought to hold promise for salt production could be managed to provide funding for the survey. He noted that to perform chemical analysis on specimens, the university/survey needed to buy lab equipment and chemicals. He observed that even though the geological survey was to have priority, if the natural history survey were done concurrently, it would save a lot of money. Lastly, ever the educator, he recommended that Minnesota's teacher colleges should receive the survey's geological specimens once the university's collection was complete.

The comprehensive, 129-page report, pulled together in the six weeks since the field season ended, was ready for President Folwell and the board of regents by December 31, 1872.

January 6, 1873

Newton Winchell puts his back to the wind as he makes his way down Main Street. His boots squeak on the snowy sidewalk. His breath comes out in frosty puffs, stiffening his scarf as it freezes and leaving delicate crystals on his whiskers. Had Ann Arbor ever been this cold? Minneapolis is nearly beyond endurance. The factories of the milling district emit billowing plumes drifting off to the south. The air is milky with ice crystals. Even the river is frozen in stretches, though not, of course, at the cataract, whose roar fills the air as night comes on.

December had been frigid. The worst cold had come on Christmas Eve day, a numbing thirty-eight degrees below zero. He and Lottie had been quite concerned taking baby Avis, Hortie, and Ima out in that weather, even to attend Christmas services.

Newton is leaving a meeting held in Dr. Asa Johnson's medical office at the corner of Main Street and Central Avenue. Eleven men from both sides of the river had gathered to form a club devoted to scientific inquiry. The majority were physicians—Newton was the only professional scientist—but most had taken sciences courses in college, although that was not a requirement. Besides the doctors, the club included a math teacher, a lawyer, a dentist, and a superintendent of schools. Its aim will be to observe nature, make collections, identify and preserve the specimens, and discuss current scientific topics. In short, the group wanted to promote high-quality scientific knowledge in the young state of Minnesota. They called their club "The Academy of Natural Sciences" and agreed to meet monthly at 3:00 p.m. at Dr. Johnson's office. They adopted the motto "Vox Naturae, Vox Dei," "The Voice of Nature Is the Voice of God."

The setting sun had streaked the winter sky with brilliant hues of coral and gold belying the frigid air. Now the blues of evening snow intensify, and Newton anticipates home, a warm kitchen, and a cheery stove crackling with fire.

The meeting had gone well. The Academy would be a good forum for his scientific ponderings, and a means by which to promote scientific understanding throughout the community.[17]

WINTER TERM AT THE UNIVERSITY OF MINNESOTA began the next day, January 7, 1873. Newton Winchell assumed duties as professor of geology that term, teaching two courses. The course catalog for that school year did not specify which courses were offered winter term, but physical geology, an entry-level course, was required for some tracks, and geology and mineralogy was required for all bachelor degrees as well as for civil engineering.

A forerunner of the university had begun, pre-statehood, in 1851 on a site in St. Anthony.[18] When the growing school needed elbow room, it moved in 1858 to a two-acre site downriver from St. Anthony Falls. A main building in the trendy Italianate style had been constructed from the bluish limestone quarried from the nearby Spirit Island quarry. It was a substantial structure, four stories high and topped with a cupola, but it had closed its doors soon after opening, the lingering effect of the 1857 financial panic.

With the onset of the Civil War in 1860, the university was mothballed, and Old Main fell into disrepair.[19] The governor appointed state senator John S. Pillsbury as a regent in 1863, and when Pillsbury inspected Old Main's condition soon after—the university was in his legislative district—he found a family squatting on the first floor and domestic turkeys penned up in the basement.[20]

Due chiefly to the persistent efforts of Pillsbury, finances of the school were righted, and a president, William Watts Folwell, called. The university was reborn. When Winchell arrived as state geologist and professor of geology, the school was set to graduate its first seniors.

Climbing the front stone steps of Old Main that January day of opening classes, Winchell looked across a snowy lawn. Here, too, slender bur oaks grew in abundance but were bereft of their leaves. Below the bluffs, the Mississippi was visible, curving under the mills hugging the riverbanks. A footbridge downriver from the falls may have also been visible.

Old Main was designed to be expanded in stages. The first stage had been the central hub, with fifty-three rooms, an assembly hall on the third floor that seated a thousand, and space on the first floor for a library, a reading room, and home bases for literary societies. After the university reclaimed its vigor, workmen began on an east wing extending from the back of the building. When Winchell opened his first class lecture, the new wing was still under construction but would be completed in a year. The natural history museum would occupy a room in the original building when space opened up.

Winchell taught five days a week for twelve weeks. Students and professors began each morning with opening devotions in the assembly hall. Besides prayers, this might include a recitation or brief talk from an upperclassman or professor. Classes followed. A full-time student carried three courses, as well as other exercises.

In the afternoon, students participated in military drills (required of all males), labs, drawing, farm and greenhouse activity for students involved in botany or agriculture, and surveying for the engineers. Winchell's students undoubtedly spent afternoons in the laboratory, working with collections of rocks. Within a few years, he was illustrating his lectures with magic lantern slides,[21] employing maps, diagrams, and models, and for the mineralogy course, identifying the chemical composition of samples using blowpipe analysis.[22]

WINCHELL'S SEASONAL SHIFT from classroom to fieldwork was interrupted in June 1873 with news of the death of his father, Horace, in Lakeville, Connecticut. Horace was seventy-two. Winchell went to Lakeville for the funeral and to help his siblings determine the future of his mother and his fifteen-year-old niece, Laura Miles, the daughter of Winchell's sister Laura, who had died at her birth. The loss sent the family matriarch, Caroline, to Milwaukee, to be cared for by Newton's brother Robby, now principal of Milwaukee High School. Newton and Lottie assumed the guardianship of Laura.

In February, Newton had visited the salt deposits at Belle Plaine that his brother Alexander had examined in June 1872. Early settlers in the region were keenly interested in possible salt deposits, chiefly for their use in preserving meat.[23] Salty water (a salt lick for deer, the Indians had said)[24] had been found at Belle Plaine, and with much hype, well diggers began drilling. State legislators conferred six sections of land to the prospectors, but the venture failed to produce much. When it asked for yet more land, the legislature requested a professional geologist. Alexander Winchell had been hired and, upon examination, had opined that the geology of the area was not favorable for salt. With Alexander's report, the governor of Minnesota was convinced it was a losing venture, but the entrepreneurs were not. More state land was granted them, and then the issue was handed to the new director of the geological survey—Newton Winchell.

At Belle Plaine, Newton discovered the drillers had dug to 710 feet, without success. He agreed with his brother and told the governor so. With that, the matter of salt was laid to rest.

A second pressing economic question the legislature wanted the survey to address concerned the state's fuel sources. The prairie region of Minnesota promised rich, fertile soil but lacked trees for building and fuel. To the south, Iowa had uncovered promising coal deposits. Might they extend into Minnesota? Iowa had also peat deposits that might reach northward.

Astute to which side his bread was buttered on, Winchell spent forty pages of his 1873 annual report discussing the nature of peat deposits in North America and Europe. He spent hours in the field examining shallow wetlands whose bottoms might be decayed sufficiently to form peat. In the end, he concluded that in southern Minnesota, where he had focused his time, peat would be more costly than wood or coal. He added there were rumors of far better peat deposits in northern Minnesota.[25]

But most of his time that first full field season in 1873 was spent moving methodically up the Minnesota River valley, examining the exposed bluffs on either side, beginning to parse the geology of southern Minnesota. Moving glaciers had left a thick blanket of till on the landscape, obscuring bedrock. River valleys, where water had cut through bedrock, and railroad cuts and quarries, where men had done the same, were good starting places to see what lay beneath.

The Minnesota River valley yawned far beyond its waterway. The placid river had evidently once been much bigger. Winchell began at Fort Snelling (Bdote), the river's confluence with the Mississippi, and worked his way southwest toward Shakopee, then to Jordan and Belle Plaine, to Ottawa, St. Peter, and Mankato. He traveled by train (for free) and, if warranted, rented a horse and covered wagon to visit outcrops and quarries. The legislature had doubled the funding to $2,000 a year, as requested, and he hired university students as assistants. They camped out most nights and took meals with nearby farmers.

That summer, he traced the different beds of the Ordovician Period, laid down long ago in the Paleozoic Era, which comprised the valley's bluffs. This was familiar work. Ohio had also revealed sedimentary rocks from the massive inland sea. He observed that while the top layer of limestone beneath the till was first visible at Shakopee, it disappeared from the bluff profile, only to reappear at Ottawa, St. Peter, and Kasota, where it was of much finer quality. There were quarries in operation in these towns. The St. Peter quarry was chiefly furnishing stone for the construction of a state hospital building.[26]

At Mankato, he discerned bluish clay in a railroad cut by a bridge. This clay was in a depression in the Ordovician limestone bluff,[27] almost as if placed there. He noticed a similar juxtaposition at the Vermillion River, at the St. Peter quarry, and at a railroad bridge over the Blue Earth River. From these various sites, he concluded that the long-weathered Ordovician limestone was covered by a Cretaceous sea in which clay was deposited, and much later buried under glacial till.[28] This interpretation was enumerated in twelve points, illustrated by drawings. Winchell's meticulous attention to important details leaps from the page in his recounting.[29]

Winchell noted that this clay when found in large enough quantities at New Ulm and Mankato could be used in pottery. The Dakota had called it "Mah-ka-to"; the French, *Terre Bleu;* and from it, Blue Earth County derives its name. This blue clay may have been the pigment that Sisseton Dakota Indians had used for ages.[30]

The Minnesota River makes a sharp bend near the town of Mankato and flows to the northeast. Upriver from Mankato at the town of Redstone, Winchell visited a quarry cutting a rosy quartzite, a metamorphic rock.[31] He was told this quartzite formed the ledge of the cascade at Sioux Falls, Dakota Territory. Gneiss outcrops appeared around the town of New Ulm and continued as he moved upriver. These rocks were below the sedimentary layers and were older. Winchell moved back in time as he made his way upriver.

August 24, 1873

Newton Winchell strides toward the slight elevation that marks the glacial ridge known as the Coteau des Prairies. The Lac qui Parle River is a thin, winding ribbon off in the distance, lined with lush willows and box elder, the only trees in sight. A stiff wind tugs at his hat. His jacket flaps in the breeze, and the tall grasses ripple like the waves of a sea. The hand of humans seems not to have touched this place, save for the railroad tracks that have brought him here. He thinks that farming would prosper in this place. Today is his wedding anniversary, and he feels expansive, he feels like writing poetry. He may attempt it tonight.

Tonight he'd write:

Nine years ago, my dear, nine years ago.
The world was all before us then.
We knew not then, my dear, what weal or woe
The nine years gone held from our ken.
Nine years ago, my dear, if we had known
Where we should be today, my dear,
Through what the joys, through what the cares
we'd come,
Should we have stayed our hands, my dear?

More verses like these and then suddenly, constrained by the difficult metric, he'd break off and exclaim, "That's too slow . . . plain prose is plainer and quicker to express even love. . . . the poetry I thought I had once has all blossomed into prose."

But poetry is ahead of him, to be written when the sun is down and the prairie sky is full of stars. Right now, as the wind blows and the grasses wave, the sky is blue and the land is endless. The future, he thinks, is promising.

FRACTURE

1873–74

The Winchells thrived in Minneapolis in fall 1873, settling into school, into work, into the neighborhood, into life. Having the room, they took in a boarder, Wesley Sawyer, a scholar of Teutonic studies, and his presence generated much-needed income for a household always just getting by.

Following the death of his father that summer, Newton had agreed to assume the care of his niece, Laura Miles, age fifteen, who had been living with her grandparents since infancy.[1] The Winchell men, Laura's uncles, had drawn up a plan for her education following the funeral. Laura moved to Minneapolis and began at Minneapolis's East District public school in the upper level, walking every day with Hortie, age eight, and Ima, age six. Newton was pleased with her academic progress and her adjustment to their home, though Laura, musical like all the Winchells, missed her piano, sold in Connecticut when the family home was dismantled.

Lottie was pregnant and due in March. Her health was good, save for the encumbrance of a growing belly. Newton was wrapping up his first full survey field season and preparing to write the annual report, his second. Still, he was bothered by nagging pain in his left leg, which had become so severe that often he slept nights in a rocking chair by the stove, where the heat gave some relief.

Newton, who was only thirty-three, had contracted the chronic pain in a simple act. He had been at the depot over a year before, and wet snow had balled up under the heel of his boot. Annoyed, he flung his foot out to clear it, and the action caused a sharp pain in the back of his knee, which over time became persistent and troublesome. He limped and favored the leg while standing. He thought that overwork during the field season had made it much worse.[2]

But the overall harmony of the Winchell household did not reflect the national mood. A financial panic in the American stock market had collapsed many banks and businesses, bringing on a depression that was felt in Canada and Europe as well. Among the many business casualties were railroads, which had been overvalued. In Minnesota, financier Jay Cooke, the backer of the Northern Pacific Railroad, had gone bankrupt, and the future of that line, once envisioned as uniting the Midwest to the Pacific coast along the northern plains, was in peril.

The Winchells' second son, Alexander Newton, was born on March 2, 1874. Lottie experienced a relatively easy labor and delivery this time. Newton went out to stable the doctor's horse after his arrival; when he came back inside, the baby had been delivered and was lying on the lounge with the blissful calm of a newborn.[3] In the family record, Newton noted the baby's blue eyes and thick hair. He added that this was the smallest baby so far: only eight and a half pounds. The little boy joined a household that now numbered eight: Besides the boarder and Laura, now sixteen, there were siblings Hortie, nine; Ima, seven; and Avis, not quite two.

One month later, Lottie had recently recovered from the birth and resumed normal activity when an epidemic of influenza seized Minneapolis, strewing widespread illness. The Winchell household was not exempt. Little Ima suffered particularly, with all the classic flu symptoms: vomiting, fever, muscle aches, and lastly, worrisome congestion. Avis, Newton, and the baby fell sick in turn, but fortunately, Lottie—the chief nurse—and the others were spared.

The illness concerned them sufficiently to call for a doctor, who arrived and administered aconite. Although aconite, also known as wolfsbane, is a toxic herb, homeopaths of the nineteenth century considered it an essential medicine and even today use it to treat influenza.[4] Dr. Goodrich observed that in ten years of practice in Minneapolis, he had never seen so many sick at the same time with the same complaint. Heretofore, an influenza epidemic had not afflicted the young city.[5] It would take months for the children to regain good health.

As spring advanced and the 1874 field season approached, Winchell faced a dilemma. The university was under construction. Two major projects—an addition to the main college and an entirely new building, the Agricultural College, housing classrooms and laboratory space—promised to more than double the school's capacity. But construction would not be complete until the fall. In the meantime, the chemical analysis of soil and rock samples stipulated by the survey could not be done. Boxes of specimens collected in the previous season remained unopened. There just wasn't room for new material.[6]

In mid-June, Winchell received an unexpected invitation that must have seemed providential. Would he consider accompanying a U.S. Army expedition to the Black Hills in the role of chief geologist? Colonel William Ludlow, engineering officer of the Department of Dakota, issued the request; General George Armstrong Custer—already a high-profile military figure in popular culture with his strawberry-blond hair and sartorial splendor (gray felt hat, buckskin shirt, red bandana at his neck)—would lead the expedition.

The attractions of the invitation were multiple. Most obviously, it offered Winchell a useful and professional way to spend a field season when he was stymied by space limitations at the university. Secondly, and not insignificantly, the pay was excellent: $450 for two months of work. His only salary from the Minnesota Geological and Natural History Survey—he was not paid additionally as a university professor—was $100 a month.[7] Thirdly, geologists thrill to the opportunity to explore unmapped terrain. His previous trip to New Mexico had whetted his appetite to see more of the West's geology. Lastly, the offer may have appealed to Winchell's sense of adventure. An 1873 expedition out of Fort Rice, Dakota Territory, had made front-page news all summer. It had followed the Yellowstone River, surveying the route of the Northern Pacific Railroad, and included vivid accounts of Indian fights. The drama and excitement may have been hard to deny, especially as the father of four now saw his future career at the geological survey set for the next twenty years.

With less than two weeks to prepare, he gathered his gear, organized equipment, took along a copy of well-known geologist F. V. Hayden's 1870 report on the geology of the Black Hills, and was on his way north to Fort Abraham Lincoln, Dakota Territory.

June 23, 1874

Newton Winchell gazes out the window, watching the Dakota prairie unfurl like a panoramic mural. The Northern Pacific gently rocks as it chuffs along, trailing a plume of smoke. He had boarded the train in Minneapolis that morning, a train headed north to Banning Junction in east-central Minnesota, and then west to Brainerd, Moorhead, and, ultimately, Bismarck, Dakota Territory, the end of the line.

The trip is an opportunity to observe a part of Minnesota he has not yet seen, and he notes with interest the sandy soil, especially around Aitkin, the extensive tracts of tamarack in low areas, stands of jack pine on the sand, and occasional outcrops of "good white pine," particularly near the Junction.

As the train crosses into Dakota Territory, he catches his first glimpse of the Red River. The river is low in June, and the floodplain, he guesses, extends for twenty-five miles west of Fargo. Beyond the floodplain, he notes the rolling prairie, the marshes dotting the landscape, the lack of a definitive divide between the watershed of the Red and that of the James River, adding "there is of course, some point . . ."

He has taken a leave of absence from the Minnesota Geological and Natural History Survey. He rationalized this would be a great opportunity to collect geological and biological specimens for the survey's embryonic natural history museum, something that couldn't be done in a normal field season. And then, this was an offer to see wild country, to be the first geologist to actually visit the Black Hills. He might never have a chance like this again.

So here he is, bound for Fort Abraham Lincoln, across the Missouri River from Bismarck, and two full months of tenting under the prairie stars.

THE 1874 EXPEDITION TO THE BLACK HILLS led by General Custer set out for the far western edge of Dakota Territory with official and unofficial mandates. Officially, it was to be an exploration of territory that had been handed to the Lakota and other bands in a treaty of 1868. That agreement had ceded all of what is now South Dakota west of the Missouri River, including the Black Hills. In return, the Indians had agreed not to harass white travelers making their way to Montana, in pursuit of gold that had been discovered there in 1865.

Unofficially, Custer and his men pursued rumors of gold in the Black Hills. Gold seekers had eyed the area for decades.[8] Residents of the little prairie towns in eastern Dakota Territory had been particularly keen on the matter. The 1868 treaty momentarily squelched the fervor, but after a few years, interest rose again. Custer personally employed two miners to accompany the expedition, and he hoped they would settle the point.[9]

The gloomy murk of the 1873 panic enveloped this unofficial mission. Discovery of gold in the Black Hills would be a welcome jump-start to a stagnant economy.[10] It also would benefit the Northern Pacific Railroad, stalled by that very depression. A gold rush promised the laying of track beyond the Missouri River, making the little communities setting down roots on the prairie very happy. In the pioneer newspapers, journalists gave voice to the general disregard for the Native American claim to the Black Hills.[11] Treaties could be changed—or broken. Congress had already entertained a bill to buy back the Black Hills from the Lakota.[12]

Custer's expedition was also to be a strong military display that would quash any thoughts of the tribes pushing the limits of this treaty. Consequently, the procession that left Fort Abraham Lincoln on July 2, 1874, was exceptionally large: 951 soldiers and officers, including ten companies of Custer's Seventh Cavalry, two companies of infantry, and sixteen brass-band musicians, riding white horses and performing nearly every evening for the general.[13]

Accompanying them were 110 wagons, each pulled by six mules; one cannon; three Gatling guns (forerunners of the modern machine gun); and sixty-one Santee and Arikara scouts, the latter from tribes who were not on good terms with the western Lakota. Custer's younger brother Tom was one of the officers, as was Frederick Grant, son of the sitting president, U. S. Grant. A second Custer brother, Boston, accompanied the train in a nonmilitary role. A St. Paul, Minnesota, photographer, William Illingworth, came along to document the undertaking, and several reporters were in tow to report for national newspapers.

Winchell was not the only scientist invited to take part. He had an assistant, A. B. Donaldson, a professor of rhetoric and English literature at the University of Minnesota. Donaldson also functioned as the official botanist of the trip, but a bad back hampered his efforts, and he collected only seventy-five species the entire fortnight.[14] Presumably because he was older, Donaldson alone was known among the expedition as "the Professor," Winchell being too young and vigorous to fit the stereotype. A young George Bird Grinnell, assistant to paleontologist O. C. Marsh of Yale University, served as naturalist and fossil specialist. Colonel Ludlow, in charge of mapmaking, served as chief engineering officer.[15]

As the expedition took leave of Fort Abraham Lincoln, Winchell straddled a dark-bay cavalry horse with two white feet and a star on its forehead. He carried an army-issued single-action Colt revolver, later known as "The Gun That Won the West." Was he aware that General Custer expected to have to fight his way into the Black Hills?[16] He didn't write that in his journal, but he did observe on the first day out that the general took "great care" to avoid a surprise attack from the Indians. Soldiers traveled ahead, ascending atop surrounding hills to deliver early warnings of encounters with Native Americans.[17] In his journal, Winchell offered no opinion whether such measures were justified.

Slowly, the lengthy caravan made its way south and west, snaking like a giant python from Fort Lincoln. It advanced only fourteen miles the first day. Summer rains had filled the sloughs; the prairie was sodden, and the laden wagons often mired. The process of setting up and breaking camp would streamline in the days to come, but seldom did the train travel more than thirty miles a day. Men and horses labored under a hot July sun. "We were all very nearly fagged out," Winchell reported at nightfall.[18] Indeed, many of the men suffered heat-induced illness that very first day.[19]

By the Fourth of July, the party had reached the North Fork of the Cannonball River in the southwestern corner of what is now North Dakota. Although traditionally the Fourth was a day off work, this was not true for the Custer expedition. The men were up early and moving by 6:00 a.m. to avoid midday heat. By 11:00 a.m., they had stopped for the day. Winchell reported a temperature of 101 degrees in the shade.

The well-known geologist Ferdinand Vandeveer Hayden is credited with being the first to describe the geology of the Black Hills region, although Hayden respected the wishes of the Lakota and had never traveled into the interior.[20] While on horseback, Winchell frequently referred to Hayden's work; he detailed the land formations, the exposed banks of rivers and buttes, and the vegetation cloaking the till deposited by the glaciers.

On July 11, nine days out, the expedition got its first views of the Black Hills. For several days, Winchell had observed lignite beds. At the Grand River, just north of the Black Hills, the bed was six feet thick. Lignite, a low-grade coal, would be mined in this vein in the future, though it would never be as desirable as other fossil fuels.

Away from the fort, the expedition had few opportunities to post letters or send dispatches to eastern newspapers. Two weeks from departure, two of the expedition's Arikara scouts took letters and dispatches to Fort Lincoln for mailing, under cover of darkness.[21] Winchell took the opportunity to write Lottie and assure her he was well. Then, the expedition turned southward.

In the far reaches of Dakota Territory, Winchell observed that the pines were short and stunted. Water, which had previously been sweet and plentiful, was scarce and sometimes alkaline. Cactus grew on the high ground. Grass for the horses was sparse.[22]

As the caravan approached the Black Hills, Winchell joined Custer, engineer Ludlow, and others to ascend Inyan Kara, a rugged mountain just northwest of the main hills, in what was at the time Wyoming Territory. The name, Lakota for "rock gatherer," gave Winchell a good deal of trouble in his journal, and he employed various spellings, always variations on the Lakota name.[23] The mountain, an elevation of several foothills with a central rock mass rising sheer up on nearly all sides, also challenged the geologist.

Winchell and his assistant, Donaldson, scrambled to the summit first, soon joined by Custer and the others. The men were rewarded with a panoramic view. However, explaining the formation had Winchell scratching his head: "as to the geology of this mountain, I am somewhat uncertain," he admitted.[24] Inyan Kara, like its neighbor Devils Tower, is an intrusion, formed when molten rock forced its way up through sedimentary rock. The process had been elucidated in *Theory of the Earth,* published in 1788 by James Hutton, but it is possible that Winchell had never encountered an intrusion of this nature before.

Winchell recorded mostly geological observations in his field journal. He wrote very little about the flamboyant, mercurial lieutenant general who commanded the thousand-man train. The reader learns nothing of the flashes of rage. By one account, one morning Tom Custer overslept, and the general set fire to the grass outside his tent; by another, Custer's temper flared at the scout, Bloody Knife, and he drew his revolver and commenced shooting.[25] Nor does Winchell write much about the deaths of several men or about the oddity of a brass band accompanying an exploratory expedition and playing Custer's signature marching tune, "Garry Owen."

On the surface, Winchell and Custer had much in common. They were nearly the same age, both born in December 1839. Both had been poor boys who had struggled to move up the social ladder; both were close to their younger brothers. Both had married Michigan farm girls, and Custer retained close ties to Monroe, Michigan, in the same corner of the state as Adrian, where the Winchells had lived as young adults.

Yet the dispositions of the two men were very different. Winchell was meticulous, detail oriented, reflective, and intellectual. Custer had finished last in his West Point graduating class and was ostentatious, even rash. A neighbor of his in-laws in Monroe observed that the Custers had "no intellectual interests [and] very little schooling."[26]

So it is perhaps not a surprise that Winchell focused on rocks as he traveled slowly on horseback through the magnificent Black Hills in late July, parsing the geology of Dakota Territory and the Black Hills. He described stands of Ponderosa pines, which he erroneously called "Norway." He identified fossils that he had seen before along the western shore of Green Bay, Lake Michigan.[27] He examined myriad rocks: a very-fined grained mica schist, a range of granite hills running north and south, a reddish fine-grained sandstone, which he designated "Minnelusa sandstone" acknowledging the Lakota name for the valley in which he first encountered it.

He broke from geologizing extensively only once, when he joined a party to climb Harney Peak, the highest elevation of the Black Hills. There were five men chosen to accompany Custer to the summit: Winchell and his assistant, Donaldson; the geographer, William Ludlow, of the Army Corps; and two others. A military escort accompanied the party halfway, and they acquiesced to the route of Ludlow, which would give him the best topographical information. As the route became more rugged, the escort gave way, and eventually, the men left horses behind to scramble on foot.

> *The views are breathtaking: "on ascending [a ridge of granite hills,] a most magnificent prospect burst suddenly upon us," he would later write, "which caused us each, on reaching the summit, to utter some exclamation of surprise and admiration." The way to the top is sometimes daunting: "trees had been thrown down by fire or tempest, often half-charred, and a thousand shrubs and small aspens . . . had made a perfect, netted mesh, through which no horse could pass." Winchell and Custer succeed together in making the final ascent. Custer takes his rifle, and Winchell grabs his hammer and barometer. Then, by painstakingly crawling up an eroded channel of the nearly vertical rock surface, they brace themselves across from side to side, and seizing every available knob of projecting feldspar, they gain purchase to the top.*
>
> *It is not quite the summit, but they reach the apex soon. The view, oh the splendid view! Nonetheless, Winchell carefully notes the "very fine and tempting array of red raspberries (Rubus strigosus) lining the way out to the summit." But then, he realizes he is exhausted and drops down to catch his breath and calm his racing pulse. Donaldson measures it for posterity at 136 beats per minute. Custer fires off three shots to salute their achievement. The party believes they are the first whites to make the ascent.*
>
> *Retracing their steps down is equally exhausting. Winchell checks his pocket watch as they drag into camp 1:20 a.m. The officers had lit bonfires to guide them to the campsite.*

WINCHELL'S MINUTE ACCOUNT of the Harney Peak climb was in striking contrast to a small item he mentioned in the same journal entry: on August 1, the miners on the expedition discovered gold and silver, and the expedition would remain in the area for several days to give the miners time to explore further.[28] He conveyed that news in fifteen words, then continued on for pages relating his adventures on Harney Peak.

This was big news, however, a game changer in the fate of the

Black Hills, the Lakota sacred site. Much later, at least one historian would pinpoint the discovery of gold as the moment at which the United States went irrevocably astray in its relationship to Native Americans.[29] Certainly, historians can draw a straight line from the discovery of gold, the breaking of the treaty with the Plains Indians in the rush that followed, and the immense gathering of tribal warriors in late June 1876 that resulted in Custer's death and the defeat of the Seventh Calvary at the Little Bighorn River.

But in this moment of discovery, in the opening hours of the prolonged drama that ensued, Winchell seemed not overly interested in the find. He didn't record hustling to the locale to verify the discovery. He didn't mention the form the gold took—dust or flecks or nuggets. He laconically added, after imparting the find, that a runner, Charley Reynolds, would travel to Fort Laramie the next day, carrying the second set of dispatches to the outside world—and trumpeting the news of gold in the Black Hills.[30]

It would seem from his spare account that Winchell did not see the verification of the presence of gold as his concern. But later, at home, in the reports he wrote up and in a talk delivered to the Minnesota Academy of Natural Sciences, he made clear that he knew his opinion would be controversial. He stated that both he and Donaldson had been shown "gold" finds in the field, and they had deemed them pyrite—fool's gold. He and others harbored suspicions that the miners had brought samples of gold with them on the expedition and then "discovered it" in their pans in the stream at the foot of Harney Peak.[31]

This report on the geology of the Black Hills, when released in September, was picked up by telegraph services and flashed all over the country. In fall 1874 Winchell became notorious—especially since the leader of the expedition, the charismatic General Custer, contradicted him. Custer, in fact, attributed Winchell's claim of not finding gold to "professional pique." The miners had not gone to Winchell, the official geologist, to confirm their find. They had merely announced it, without Winchell's professional sanction.[32] The general stated baldly to the press, "Why Professor Winchell saw no gold was simply due to the fact that he neglected to look for it."[33]

That the chief geologist and the expedition head disagreed about the presence of Black Hills gold meant that the government would send another expedition the following year, 1875, with a different geologist and a different military leader. That geologist, Walter P. Jenny, of the Columbia School of Mines in Washington, D.C., did find gold in enough quantity that would make mining profitable.

That conclusion alone would have caused consternation in Winchell. He prided himself on his professional expertise. How humiliating to be defamed in the national press. The Bismarck *Tribune,* which had the biggest monetary stake in a Black Hills gold rush, brayed the loudest: Winchell was "an ass and a Dogberry"; "As a geologist he is rather thin."[34]

The debacle can partly be explained by Winchell's stringent code of conduct: he only confirmed findings that he, himself, had seen personally. But that doesn't explain why he apparently didn't regard the occurrence of precious metals as his domain. In addition, Winchell seems to have been uncharacteristically tone-deaf to the political underpinnings of the expedition.

Politics and professional pride aside, the experience yielded some benefit to Winchell and the University of Minnesota. He shipped home eight antelope skins, one elk skin and head, one bear skin (unspecified species), and one weasel skin, all for the young natural history museum.[35]

WHEN NEWTON WINCHELL ARRIVED HOME from Fort Abraham Lincoln in late August 1874, he was greeted by the frightening, convulsive gasps of whooping cough. His youngest two children had contracted the dangerous disease in July and had been ill for weeks. Lottie, herself, was suffering from nervous exhaustion. Still nursing the new baby, she had lost weight from chronic diarrhea.[36] Winchell described her as "skin and bones." The responsibility of sick children without a spouse at home to share the worry and the care consumed her. Increasing tension between her and the teenage Laura, who told her cousins that "she was not obliged to mind 'Aunt Lottie or Uncle Newton,'"[37] nearly did Lottie in.

Winchell had planned to head immediately to southern Minnesota for the survey's abbreviated field season, but instead stayed in Minneapolis and attempted to right the household. Lottie was sent off to recuperate in Michigan with friends, taking the three youngest children and a nurse with her. She stayed there until December 18. Laura and Hortie, who needed to be at school, remained at home, while Winchell swung in and out of Minneapolis, surveying Freeborn and Mower Counties.

The family reunited for Christmas and had a brief respite from illness. But then another childhood disease invaded the home. Avis and Alexander recovered, finally, from whooping cough, when even before the Christmas season ended, Ima and Avis contracted scarlet fever.

Newton and Lottie dreaded this disease in particular. It killed so many children and damaged hearts and kidneys. Years before, when they were neighbors of the Stephensons in Adrian, Michigan, Lottie had asked Dr. Stephenson how to treat scarlet fever and had written down the instructions. Now, she followed them exactly. The girls were immediately quarantined, and Hortie strictly forbidden to come near the sickroom. The incorrigible ten-year-old did not heed the stricture, however, and he and Alex caught it. The beset parents tended the sickroom day and night, even as winter term at the university began and Newton embarked on his teaching responsibilities.

After a brief interval of health in late winter, chicken pox made the rounds at the elementary school. One by one, the Winchell children broke out and were confined again to the sickroom.

Growing strife with Laura heightened the stress of continually sick children. The elder Winchells had no experience with teenagers. Laura was unhappy in school and wanted to quit. She missed her music and her grandmother in Milwaukee. Newton observed that whenever Laura received a letter from Grandma, another round of conflict ensued.[38]

When Grandma Caroline learned that her own mother, over eighty years, in New York State was failing, she wanted Laura to drop out of school and accompany her home. Here was the out Laura had longed for, but Newton was adamant: he would follow the plan that he and his brothers had laid out for their niece back in Lakeville when her grandfather had died. Laura needed to complete high school and then could take a specialized year of music.[39] Unbeknownst to Newton, Laura began to look for work in the only field she knew: making hats.

The domestic drama was compounded by romance. Laura had fallen in love with William Peck, the brother-in-law of Stephen Peckham, Winchell's colleague at the university and the chemist on the geological survey. Stephen and Mary Peckham were also neighbors to the Winchells and privy to the increasing familial conflict.

In matters of family disagreement or financial need, Grandma Caroline tended to appeal to her oldest son, Alexander, to intercede and provide what was needed—money, counsel, mediation, whatever would fix the problem.[40] Now Alexander was drawn into the conflict in Newton's household.

One day, after a spectacular blowout, Laura left the home and found refuge with the Peckhams. Newton declared she could not return to their family unless she humbly apologized and pledged good behavior.[41] When in a letter, Alexander suggested his younger brother had driven Laura out of the house, Newton hotly denied it, adding, "I had expected that turn would be given the movement." Newton claimed to have overlooked many disobedient acts and impudent remarks. She had tried to pit him against Lottie. She had taught the children bad language and bad manners. Her presence had been a constant source of discord in a family that had formerly been harmonious. Newton had in great embarrassment put up with it for a very long time. "I do not believe," he added, "that you or Rob would have endured what I have." He ended emphatically, "I say I am done."[42]

Grandma was distraught and furious at her son—and daughter-in-law Lottie, whom she blamed even more. "It is most shameful," she wrote to Alexander, "to have N. turn her out this way, even if she was greatly to blame, he ought to have sheltered and protected her as he agreed to do in Lakeville."[43]

The family turmoil swirled as Newton prepared for the survey's 1875 field season. Then, in July, seventeen-year-old Laura married William Peck at the county courthouse, the Peckhams serving as witnesses. The couple settled into a house in the same neighborhood as the Peckhams and the Winchells, and that chapter of family conflict seemed somewhat resolved.

Lottie and Newton, for their part, had of necessity turned their attention to their small children, since sixteen-month-old Alexander was now gravely ill with dysentery. Newton, who had just gotten into the field, interrupted the work to make an emergency trip home when Lottie and the doctor deemed the baby near death.[44] By later in July, little Alex seemed to have made it through the worst of it, and the parents looked forward to a slow but steady recovery. Perhaps the fractured time was drawing to a close.

BEDROCK AND RIVERS

1875–78

THE GEOLOGICAL AND NATURAL HISTORY SURVEY began to blossom under Newton Winchell's guidance. Winchell bloomed as well, as he recognized the wide-ranging possibilities of such a scientific endeavor. In the first decade of the survey, Winchell claimed for the survey—and himself—the tasks of investigating and removing the varied roadblocks that hindered settlement of young Minnesota. At the same time, Winchell laid the foundations for what one historian would call "one of the most complete and carefully detailed projects of the kind in American scientific history."[1]

From 1875 to 1878, the survey reached all four corners of Minnesota, from the dry prairies of Rock and Pipestone Counties to the sculptured limestone bluffs of Houston County; from the yawning expanse of the Red River valley to the wild, untouched forests of the Arrowhead region between Lake Superior and Ontario.

Winchell hired help as needed, displaying a keen eye for talent and launching future scientists, especially geologists. This keen eye would make him famous as one after another budding geologist cut their professional teeth under his guidance. The survey took on the pressing issues facing the young state: the stability of St. Anthony Falls, the driving force of its major city; the grasshopper infestation that threatened the permanence of agriculture in western Minnesota; and the location of ores in the northeast. Winchell would mine these practical questions for theoretical significance and establish himself internationally as a geologist of heft.

In 1874, Newton and Lottie moved into a home south of the University of Minnesota campus. They remained there for the next three decades. The house at 120 State Street was originally part of the Cheever Hotel, an early pioneer hostelry built by William Cheever on land overlooking the river.[2] Cheever also operated a ferry directly below the hotel, and an observation tower that one historian has called Minnesota's first tourist trap.[3] From the tower, visitors could overlook the Mississippi, the raw streets of Minneapolis, and the shimmering chain of lakes beyond. A sign urged sightseers to "Pay your dime and climb."

Winchell enlisted the help of Mitchell Rhame, university civil engineering professor, and Wesley Sawyer to move the house five hundred yards to the State Street lot. The three academics intended to eventually offer a boarding house for university women students, an idea proposed by Lottie, who had noticed that although the University of Minnesota claimed to be coeducational, there were very limited living options for women unable to live at home.[4]

The former hotel had eighteen rooms, ample space for the growing Winchell brood. A frame structure with clapboard siding, the house was L-shaped with a porch, a bay window, and curlicue Italianate brackets adorning the corners. The Winchells seeded a garden, dug a root cellar, built a barn, and planted trees. As the saplings grew, Newton, when he was away on survey work, wrote Lottie instructing her to see that the delivery horse didn't nip the cottonwood, and to water the elms before they showed signs of stress.

A succession of renters called the place home immediately. Often workers on the survey stayed there, and later, students took rooms. As the Winchell children matured into university students themselves, 120 State Street became a hub for student social life.

IN 1875, WINCHELL HIRED MARK W. HARRINGTON, a University of Michigan graduate and trained botanist, formerly of the Michigan Geological Survey, to survey Olmsted, Steele, and Dodge Counties.[5] Winchell went to Fillmore County, on the state's southeastern border, which had substantial rock outcroppings. University student William Leonard accompanied him. Leonard had reluctantly taken Winchell's mineralogy course, a requirement, but soon warmed to the subject. Now the pair traveled Fillmore County by horse and platform wagon, which Winchell called, grandly, "the state wagon."[6] To pass the time between geological sites, the two took turns reading aloud from Dickens's *Great Expectations.*

Fillmore County greatly interested Winchell. Drained almost entirely by the Root River, the county was dissected by narrow, deeply cut valleys with towering, rocky bluffs. The same limestone and sandstone beds that Winchell had seen along the Minnesota River were also visible in this county. The Root River had cut deep channels through the beds, exposing the layers. At certain points along the river and its many tributaries, where hard limestone forming the riverbed gave way to softer sandstone, the sandstone had eroded, creating a drop and resulting in waterfalls or rapids. Waterfalls, even modest ones, could be harnessed for power, and settlements in Fillmore County—Preston, Chatfield, Forestville, and Lanesboro—all began at these falls.

Winchell's geological eye saw the presence of mills as evidence that a waterway's bedrock was changing from sandstone to limestone. The border of the limestone bed was at that point. These "knickpoints"—the points at which the slope of a riverbed changes—could then be connected to map the various layers of bedrock underlying the county. As he explained his reasoning to the university regents in his 1875 annual report, Winchell noted that the Falls of St. Anthony also formed that way.[7]

To identify the several limestone and sandstone beds of the exposed rock, Winchell relied not only on their appearance and location in relationship to each other but also on the numerous fossils found within a bed. He knew which were common and which were not, and that the fossils that many residents considered "little petrified turtles" were actually brachiopods.[8] He identified the cephalopod *Orthoceras,* which quarrymen encountered at work and thought were big snake fossils, and *Chaetetes,* a fossil coral, collecting and labeling and carefully packing all for shipping to the lab in Minneapolis.

Fillmore County also intrigued Winchell because it lay on the western edge of what had earlier been identified as the "driftless area," that part of the continent's midsection that lacked evidence of glaciation. As glaciers move along the earth's surface, they pick up large amounts of debris—rocks and soil—and deposit it at the ice margins as they melt and recede. The driftless area lacked this glacial till.

To Winchell, the boundary between glaciated and nonglaciated land was not well defined. Even in what was purported to be driftless, there was material—small gravel and stones—that seemed to have a glacial origination. Furthermore, the till he detected in the eastern part of Fillmore County, part of the "driftless area," was different than the till in the western part of the county, which resembled the till in western Minnesota. The eastern till appeared to be older. Winchell concluded the area had experienced at least two periods of glaciation, separated by an interglacial period, which had been warm enough that peat moss and coniferous trees had been able to grow.[9] People drilling wells sometimes encountered peaty vegetation within a bed of till deposits.

Winchell was adding to geology's general understanding of the glaciation of North America as he was going about the practical work of evaluating the land for agriculture, brick making, quarrying, and possible fuel sources. In regards to the latter, Winchell noted that Fillmore County would need to rely on its wood for fuel, having no other economically feasible fuel. Trees, however, were increasing in the county, as white settlers suppressed natural fire. He saw thousands of acres of young trees not more than five or six inches in diameter on land that had once been prairie.[10]

While Winchell examined river bluffs, visited quarries, and pondered railroad cuts in Fillmore County, most of the counties in the southwestern corner of Minnesota dealt with a problem that seemed biblical. For the past three summers, hordes of voracious locusts had swarmed over farm fields, devouring every living plant they encountered. The insects, which settlers termed grasshoppers, had first appeared in 1873. They crossed the Dakota border and alit on growing crops, which they promptly consumed. Farm families frequently lost their entire summer's income. Hunger and want threatened the very existence of agriculture in the young state.

The locusts laid eggs that hatched the next summer, and the insects pushed eastward toward Mankato and north to Hutchinson. Settlers implored their government to do something. The state legislature's impulse was to convene a special commission to study the phenomenon, but in his 1875 fourth annual survey report, Winchell pointed out that such an investigation was rightly the domain of the survey.[11] If permitted to take on this entomological phenomenon, he, as survey head, could ensure good science, and the state would avoid duplication of effort. The next annual report, issued in 1876—when the locust invasion was at its most far-ranging—contained a lengthy description of, if not a scientific explanation for, the infestation.[12]

After inflicting more damage in 1877, the locusts disappeared, never to return in such large numbers. Termed today the Rocky Mountain locust, *Melanoplus spretus* went extinct in the decades following the 1870s.[13] Scientists are not certain what brought about the extinction. Locusts swarm only under certain environmental conditions. Drought in the early 1870s encouraged the insects to swarm; a wet cycle beginning in the late 1870s brought about a nonmigratory form of the insect, which was thought to occupy a very small area on the Great Plains. Cultivation of farm fields is thought to have destroyed this habitat, causing extinction.[14]

DURING NEWTON'S LONG ABSENCES, Lottie presided over a household that became more complex as the children grew. She took on more renters and expanded her activities into the public realm. Hortie, in particular, was both a joy and a handful. By age ten he listed in a rounded, childish hand all the books he had read. Some, like nursery rhymes and *Pictures and Stories of Animals,* were likely read by many children, but the list also included *Great Expectations* and *Our Mutual Friend* by Charles Dickens, and William Shakespeare's plays *King Richard II, Julius Caesar, King Henry IV,* and *The Tempest,* as well as the boyish adventures of Robinson Crusoe and Swiss Family Robinson.[15]

His schoolteacher parents recognized a child needing stimulation, and perhaps for this reason and to get him out of his mother's hair ("Hortie especially can be of great aid to you, or a great trial and trouble," Newton observed in a letter to Lottie),[16] Hortie accompanied Newton on his survey of Houston County in July 1876. Hortie kept a journal of his summer adventure but refused to write to his mother.[17] While Newton inspected rock outcroppings, Hortie climbed trees and collected an assortment of eggs and nests.[18] He got invitations to croquet (Hortie reported that he "whitewashed" his opponent) and to a local Sunday school.

Newton had been nurtured first by his own father and later by Alexander, and he had guided his brothers Robby and Charley. Now, he began another phase of Winchell mentoring as he ushered the next generation into the geological world.

Winchell first became interested in the Quaternary Period, a time of successive ice ages, during his years on the Michigan survey. As a field geologist barely into his thirties, he had pondered the movement of great ice sheets, how existing bedrock might deflect their flow, and how rivers draining large bodies of water could change course, flowing in one direction and then in another.

Hennepin County, Minneapolis's county, was routinely surveyed as one county among seventy-six,[19] but Winchell's work there had added significance since it contained the Falls of St. Anthony, the only waterfall on the Mississippi River. The underlying hard, erosion-resistant limestone lay atop very soft sandstone. At the edge of the limestone layer, the sandstone was exposed and eroded away, creating a ledge over which the river spilled.

N. H. Winchell pauses before an excavation cutting through the earth at the corner of Washington Avenue North and Sixth Avenue North, upriver from Bridge Square, Minneapolis's main business district. Along the river, sawmills hug the banks, mills turning white pine into lumber and shingles; mills providing products for the expanding city. To his back, toward town, newly laid railroad tracks run through both freight and passenger depots along Fourth Avenue North. The north-side neighborhood teems with life as people bustle, running errands, meeting to discuss business deals, exchanging news, passing tidbits of information. The small-scale buildings—hotels, taverns, some stores, liveries—are mostly frame and not long for this world. Minneapolis is booming and bigger, more substantial brick buildings will replace the wooden structures.

The exposed earth is one sample of many that Winchell will scrutinize as he works out the geology of Hennepin County. Fingering the top layer of material, a rich dark loam, he sees that is has pea-sized calcium-containing particles. This loam is two to four feet thick. Beneath it is a fine-particled clay, what Winchell deems a "brick-clay." It is yellow, as is the clay used by the Union Brick Company nearby. Other brickyards in the neighborhood are also turning out yellow bricks, the material for many of the brick buildings going up in Minneapolis. A block away from the intersection, this same clay layer contains bits of shells belonging to freshwater mollusks. This confirms the area had once been submerged.

The intersection lies in the valley of Bassett Creek. Winchell finds the clay throughout the creek valley. It lies on top of fine, white St. Peter Sandstone, which forms the bedrock in this area of the city. Farther south along the river, the waterfalls spill over a hard ledge of limestone, but here in north Minneapolis, where the slope of the land descends to Bassett Creek, the limestone has worn away. This fact supports his carefully constructed story of how the last glacier affected Hennepin County, where St. Anthony Falls once was, and why it has changed location in the past few centuries.

Winchell fishes his tape measure, a pencil, and his well-worn field notebook out of his pockets and begins to scribble notes.

St. Anthony Falls had become the powerhouse of Minneapolis, which harnessed it for work. Dozens of mills sprang up in the city's first thirty years, and commerce rapidly expanded, yet the business community was anxious. In 1869, entrepreneurs attempted to tunnel under the falls via the soft sandstone to divert yet more water to power yet another mill. The ledge started to crumble. Residents feared the falls would collapse, changing instantly from a waterfall to a mere rapids and denying Minneapolis its power source.

Businessmen living in St. Anthony since the 1850s had witnessed the falls retreating upriver as piece by piece the limestone ledge broke off. Dr. A. E. Johnson, a founding member of the Minnesota Academy of Natural Sciences, whom Winchell considered a close observer; Sumner Farnham, who with his partner had owned a sawmill on Hennepin Island that had once been immediately at the foot of the falls; and brothers Samuel and Richard Chute, who sat on the board of the St. Anthony Falls Water Power Company, could all attest to drastic recession in twenty years' time.

In 1869 after the tunnel fiasco, when the falls appeared to be on the verge of collapse, citizens on both sides of the Mississippi River sought a permanent fix to save the cataract. A lasting solution and the money to pay for it was slow in coming. City councils squabbled and dithered, and politics suffused all proposed solutions. City officials floated a proposal to issue bonds to finance the project, but citizens became incensed over the idea of taking on public debt to save private companies. In the end, the federal government stepped in with funding and expertise. The U.S. Army Corps of Engineers built a permanent apron, completed in 1880, to protect the limestone ledge, thus creating a smooth, industrial-looking curtain where once pounded a wild waterfall; drove a massive underground dike upriver from the apron to prevent cavities from forming in the soft sandstone that might further threaten the limestone; and built two low dams upriver to regulate a safe water level over the apron.[20]

Winchell devoted his remaining months in the 1876 field season to the waterfall thus altered and to Hennepin County's geology. In his fifth annual report, he informed the board of regents and legislators who waded through the 247-page tome that his survey work established these scientifically important facts: first, that the glacial till in the county was of two distinct types, one red in color, containing iron compounds, and one gray. Generally, the red till was thicker in the eastern part of the county, the gray till more prominent in the western part of the county and always overlaying it, which meant it was more recently deposited. From this he inferred there had been at least two different glaciers, one arriving first from the east, the second arriving later from the northwest.

Second, in areas of Hennepin County, clay was deposited atop the till by water action. Atop the clay, loamy soil formed as colonizing plants invaded the bare surface, and their decayed remains mixed with the glacial till, although Winchell did not mention the role of plants in soil formation in 1876.[21]

Third, concerning the Mississippi and Minnesota Rivers, which lay at the heart of white settlement in Minnesota, Winchell proposed this: when the last glacier melted, its vast waters drained through Big Stone Lake and Lake Traverse in western Minnesota through the channel of the Minnesota River. That river had clearly once carried much more water—look at how wide its valley was. This idea was not new. It had been proposed by Gen. G. K. Warren in 1868 when he served with the U.S. Army Corps of Engineers, surveying railroad routes through the Midwest. Below the point where the Minnesota River and the Mississippi River unite at Fort Snelling, the river channel of the combined rivers resembles that of the Minnesota River, with a very wide valley. The valley is evidence that it too was carved by the ancient river that was a major drainage of what is now called Glacial Lake Agassiz, between 11,700 and 9,400 radiocarbon years ago.[22] This ancient river is now called the River Warren, after General Warren.

Winchell also proposed that prior to the last glaciation, the Mississippi River did not flow through the dramatic gorge it now does. Through his examination of various brickyards, cuts, and quarries, Winchell knew that the top limestone ledge was not a continuous sheet under Minneapolis. He proposed that the Mississippi once flowed around the western edge of limestone, taking a path of least resistance, through the pronounced valley of Bassett Creek and Brownie, Cedar, Calhoun (Bde Maka Ska), and Harriet Lakes. What is now a chain of lakes was once the river channel cut into the limestone and sandstone bedrock. This ancient channel was subsequently covered up by glacial till. The ancient "Mississippi" then joined the Minnesota at some point between Shakopee and Fort Snelling.[23]

When the most recent glacial ice in the area (what is now called the Des Moines Lobe) approached from the northwest, a secondary lobe of ice (today called the Grantsburg Sublobe) moved northeast into the area of the Twin Cities and deposited more till, which choked the valley of the ancient Mississippi River and forced the river eastward to run over the limestone layer. Standing water pooled. Fresh off the glacier, it would have carried sediment. As the river slowed when it reached nearly level land, its sediment settled out atop the till, forming the clay layer, which would later be used for brick. The Mississippi, now meandering out of its original channel, continued to rise and then to plunge over the limestone edge, meeting the now enormous Minnesota River at about Fort Snelling. This is where St. Anthony Falls began.

Winchell lacked a means of dating geological events. Use of radioactive decay as an "atomic clock" had not yet been discovered. Still, if he assumed the falls had begun at Fort Snelling during the retreat of the last glaciation, he could measure the distance between there and the present position of the falls. Calculating the rate of recession, he would be able to estimate the time since the last glacier, since rate = distance/time.

He turned to detailed accounts of the falls that the earliest explorers wrote. There were many: Father Hennepin had visited the falls in 1680 and left the first written account. Jonathan Carver had seen them in 1766; Zebulon Pike had written a description in 1805; Stephen Long, in 1817 and in 1823; G. W. Featherstonhaugh, in 1835; and lastly, he had the eyewitness accounts of living people from 1856. Permanent mills were erected after that date, and this human tampering changed the rate of recession. Reports after 1856 were useless in calculating the natural rate.

Historically, there were six islands in the river, whose positions were fixed but whose relationships to the falls changed as the cataract receded. Earlier explorers all noted the islands, and from this, Winchell calculated the rate of recession between Hennepin's first account in 1680 to the position in 1856 as 5.15 feet per year. Fort Snelling was about eight miles from the falls. From this, Winchell calculated that the falls were last at Fort Snelling 8,202 years before, which would have been the date when the last glacier receded in this region.

Winchell concluded that the last glaciation period ended about 8,000 years ago, a length of time much more recent than others had thought.[24] He defended his calculation as based on data more accurate than had been used in previous attempts.

Winchell's estimate of the age of the last glaciation and his calculation of the rate of recession of the Falls of St. Anthony would prove to be one of his most influential contributions to geology. The estimate is remarkably accurate. Geologists now gauge the last glacial retreat at about 12,000 radiocarbon years ago.[25] His calculations were first published in the fifth annual report and later received international recognition when he prepared a paper that was read before the Geological Society of London at their annual meeting in 1878.[26]

Following publication of his work, geologists calculated the rate of recession of Niagara Falls, something that had not been attempted, and the estimate agreed with Winchell's estimate for the Falls of St. Anthony.[27]

At the end of Winchell's life and in the decades to come, geologists would remember this feat of clear-eyed reasoning and simple calculation. He would be lauded in memorials,[28] and a footpath running in the gorge on the west side of the Mississippi River in south Minneapolis today bears his name. After Newton's death, Hortie, grown to an influential geologist in his own right, would place a boulder from the Mesabi Iron Range at the head of the Winchell Trail on Franklin Avenue where it crosses West River Road.

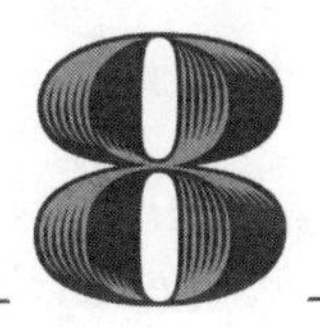

THE WINCHELLS BROADEN THEIR REACH

1876–79

THE UNIVERSITY'S GEOLOGY AND Natural History Museum occupied two sunny, light-filled rooms on the third floor of Old Main, an imposing limestone building dominating the campus. One room exhibited rocks and minerals in custom-made cases. The other housed several large mammals (an elk, an antelope, a white-tailed deer) that a taxidermist had stuffed and mounted. In addition, this room displayed plaster casts of various animal fossils, like brachiopods, sea lilies (crinoids), and trilobites crafted in the lab of Henry A. Ward of Rochester, New York.

Natural history museums were gaining in popularity in the later 1800s, but many small schools could not afford to buy or collect genuine remains of ancient organisms. Plaster casts offered reasonable facsimiles of rare and common fossils. Ward sold to high-profile museums, and he often provided avenues by which Europeans could sell their models in the United States. Leopold and Rudolf Blasschka, famous today for their finely wrought glass flowers, owned by Harvard, partnered with Ward.[1]

Next door to the museum was the mineralogy lab, where students could handle specimens, observing key identifying features, testing for hardness and reactivity with certain solutions. Science students had access to the museum from the lab or after class to ponder the displays, learning classification, distribution, and anatomy of mammals.

Students trudged the steps to the third floor every day for opening chapel services in the thousand-seat auditorium just down the hall from the museum. It would have been easy for students to wander into the rooms for a few minutes after chapel, but this didn't happen, because the museum was locked in 1877 and had been for months, while the director, Newton Winchell, remained preoccupied with other duties.

Must the museum be closed continually? the student newspaper queried peevishly. Students could peer through a window and see a cast of *Megatherium,* an extinct ground sloth the size of an elephant, lying arranged on the floor. Winchell had laid out the skeleton in pieces when it arrived, but hadn't found time to erect the form. It had lain in place for over a year. That hardly seemed sufficient, sniffed the newspaper, to warrant closing the museum for an entire term.[2]

Winchell now had funds to hire a museum assistant, Clarence Herrick, an eighteen-year-old with a good mind and a keen interest in science. In his first year on the job, Herrick added one hundred local birds to the museum's collection and began identifying and labeling the fossils collected from past seasons.[3] Herrick would publish a noteworthy study of Minnesota freshwater invertebrates in a future survey report, complete with original drawings.[4] But even with an assistant, Winchell had little time to curate a museum.

In October 1876, Winchell took a month away from his fieldwork to visit Philadelphia. The extravagant Centennial Exposition, the nation's first world's fair, was drawing to a close, and Winchell was keen to purchase materials from dismantled displays to add to the museum stock. He had stiff competition. The Smithsonian and other prestigious institutions had snapped up almost everything of value.[5] Nonetheless, he rented a room in town, printed cards requesting material "illustrative of Geology, Mineralogy or Zoology," and roamed the exposition. He collected six packing boxes of specimens and shipped them back to Minneapolis.

The immense exposition overwhelmed Winchell. He was at a loss for words to describe it to Lottie, writing only that he had seen the five main buildings in three days.[6] The fair, celebrating the United States' hundredth birthday, had two hundred structures displaying the era's technological marvels, including sewing machines, typewriters, and Alexander Graham Bell's telephone. Visitors could also buy new products such as Heinz Ketchup and Hires Root Beer and ponder the usefulness of the recently imported kudzu vine in controlling soil erosion.

Winchell ventured beyond the fairgrounds and saw Independence Hall, the Liberty Bell, and the original copy of the Declaration of Independence, kept in a large, iron fireproof safe when not on display. He wrote Lottie that he wished every day that she could be with him to sightsee and enjoy the travel.[7] He also had messages for each child: Hortie should tend the yard, Ima should not talk all the time, Avis should not be snappish, and little Alex, age two, who must have just learned to talk, should remember to speak quietly in the house.[8]

AT HOME, LOTTIE HAD EMBARKED on her own adventure. In 1875, Minnesota adopted a constitutional amendment allowing women to vote on public education issues and to hold office on school boards. The 1876 legislature ratified the amendment, and within days two men associated with the East Minneapolis schools nominated Lottie and another woman, Charlotte Van Cleve, to run for the district school board.

Prior to the school board election, women gathered to highlight the historic occasion. Many rose to deliver impromptu remarks. Mahala Fisk Pillsbury, wife of the sitting governor, John S. Pillsbury, was one.[9] Lottie, an ardent proponent of women's suffrage, was present and later wrote, "Ladies, holding a political meeting! A caucus! Yes, really, a political meeting, their first, too." Later, she noted, "Women of the state have accepted the trust in good faith and have actually—don't be shocked, now—gone to the polls and voted."[10]

Despite an April snowstorm, the women of East Minneapolis turned out to vote, sometimes accompanied by their husbands, some alone, others in small groups. Both women candidates prevailed by an ample majority. Lottie and Charlotte Van Cleve took their places on the school board the next week.

Lottie carefully clipped and saved a local newspaper account of the meeting. Describing her as a "lively, active, pleasant little woman," the *Evening Mail* noted her merry laugh and rapid speech. The *Mail* reported that Mrs. Winchell asked for a copy of the school charter and then wanted to know if the school had a library. Also, whether court fines went to fund it—that was true for some communities.

Mrs. Van Cleve asked if students had access to the library's books. Or was it only for teachers? The board elected Lottie as secretary. She immediately began compiling a history of the public schools in East Minneapolis.[11]

The *Evening Mail* commented that both women wore black dresses trimmed with velvet. Lottie wore a red bow around her neck. Mrs. Van Cleve wore a white lace scarf with a white coral pin. Both women wore hats and gloves, which they removed and placed on the table before the meeting. After, a board director, Mr. Smith, escorted the women to their carriage, unfastened the horse, and bade them goodnight.[12]

Following the election from his home in Milwaukee, Rob Winchell sent his sister-in-law a postcard of congratulations. He was pleased with the results, and he added, "The world moves."[13]

WHILE LOTTIE MADE HEADLINES in the local papers, Newton continued his teaching obligations at the university. Winter term he had taught dynamical geology and a mineral and lithology course. Spring term he lectured in historical geology, botany, and zoology. He was an engaging professor with his keen intellect, and cut a notable figure on campus with his big, bushy beard. Jokes about geology found their way into the student newspaper:

Professor of Mineralogy: "What is the difference between Hematite and Magnetite?"

Junior: "I think one contains no water."

Professor: "That's right. Neither does the other."[14]

and

Junior in Blow Pipe Analysis: "Professor, have I got Apatite?"

Professor, absent-mindedly: "I guess so. You generally have!"[15]

ONE JANUARY DAY IN 1878, after a lecture on glacial erratics, out-of-place rocks transported to locales by glaciers, Winchell strode into his lecture hall to find a huge boulder, several hundred pounds, resting on two chairs in front of his table. Twenty enthusiastic geology students had hauled it up to the third-floor lecture room via the elevator. It bore "striations," demonstrating its glacial affinity—duly scratched in with a file. The professor nodded and complimented his class on its perspicacity.[16] It was, he declared, an excellent specimen.[17]

The stone was immediately famous and ultimately esteemed. The graduating class that year designated it as its "memorial stone." The students likened themselves to the transported rock: formed under pressure, they had in their four years at the university been slowly ground down. They placed the "erratic" at the base of an elm, for posterity.[18]

But the teaching proved to be too much for Winchell. He simply did not have the time necessary to teach five courses and work on the survey. In spring 1879, the university hired Christopher W. Hall, a graduate of Middlebury College in Vermont and a former school principal at Owatonna, to teach Winchell's courses and later assigned Hall the winter geology classes as well. Hall assumed the title assistant zoology professor and was appointed as assistant state geologist.[19] He was a well-rounded naturalist who very quickly would prove his usefulness to Winchell and the university.

Relieved of teaching duties, Winchell remained a dominant and articulate voice for science and for public education. In January 1880, after serving a year as the president of the Minnesota Academy of Natural Sciences, Winchell delivered a parting speech that was a ringing defense of the scientific endeavor. The academy must have shown signs of faltering as an organization, because Winchell exhorted the seven-year-old society to persevere. Using biblical language, he called upon the academy to "arise and shine": the organization was at a crossroads where it might "die ingloriously." "Let [the end] not come upon us unawares."[20]

Science, he claimed, was the great achievement of the nineteenth century. "Our literature, our thoughts, our lives, are permeated with its influence. Our language takes on its terms of expression, and the very changes which [the language] experiences . . . are explained and justified on 'scientific' grounds."[21]

Winchell was well aware of how his own profession, geology, had changed the intellectual worldview in a mere century. Scientists now thought the earth was very old and constantly changing, and that had replaced the former belief of a young, created planet that remained static. "The progress of science in the past hundred years has reacted upon other branches of knowledge," he averred. "If it has overturned some it has established others on surer foundations. It has everywhere been victorious."[22]

Winchell pointed out that Minnesota was a wonderful natural laboratory for scientific endeavor. The Mississippi River, the Red River of the North, and Lake Superior—"the greatest body of fresh water on the globe"—all offered opportunities to study hydraulics, drainage and flooding, bird migrations, and geology, subjects whose surfaces had not even been scratched.

Of what practical good would this pursuit of science be? "There is a very close connection between immigration, settlement, wealth, and the prosecution of science." He noted that scientific knowledge fosters technological invention; invention spawns development, which attracts skilled labor, through immigration. Furthermore, science encourages the kind of citizens most beneficial to society: "Science provokes a quick and observing eye. She requires the cool and steady judgment. She skills the hand to its gentlest and nicest touch. She makes us tolerant of opposition and willing to be corrected. She would harmonize our disagreements. She would systematize our efforts; she would regulate our pleasures, and she would enhance the happiness of our homes."[23]

Revealing more than a whiff of natural theology—the concept that God reveals Himself in nature—Winchell added that science enhances the intellectual and moral good, and an academy of science promotes scientific endeavor. "Whatever leads us into the mysteries of nature, whatever introduces us to the great plans and unsearchable wisdom of the Creator . . . [these all] conduct us 'from nature up to nature's God.'"[24]

Still, Winchell had very firm ideas on which institution should be in charge of the scientific pursuit: not the church. A year after delivering his passionate farewell as academy president, he was once more at its podium, having again been elected by the membership for a one-year term. This time, his topic was "The State and Higher Education."[25] The president of Hamline University, David Clarke John, had stated that the government should not be in the business of funding colleges and universities. Higher education was the domain of the church, which had been performing the task better and longer than the land-grant institutions that were established by an act of Congress and were a "conspicuous and universally acknowledged failure."[26]

Winchell possessed the temperament to rise to the bait, and he was well suited to carry the banner into battle. He was a practicing

Methodist, active in the leadership of First Methodist–Episcopal Church in East Minneapolis, and its official church historian. Hamline was affiliated with the Methodist Church, and President John was an ordained clergyman.

Winchell began his defense with a schoolteacher's overview of history, of how the Reformation had established "the right of every human mind to think, and to be responsible for its own acts," and this carried with it "the necessity of becoming well-informed."[27] Americans had a duty to be educated, and in a democracy where the people *are* the government, state-sponsored universities are actually the people educating themselves. This contrasted with England, where the first universities were formed by the church to educate clergy.

Using statistics, Winchell examined the charge that state schools had failed. These land-grant schools taught subjects spelled out in legislation: engineering, agriculture, mechanic arts, and other fields with immediate, practical value. Winchell thought that these fields actually popularized higher education, making it available to many young adults and not just a wealthy few. He thought, in fact, that the state schools contributed not only to the spread of science but "also to the spread of Christianity, particularly among those intelligent classes of the people who have been hostile to it, or indifferent, because of the attitude of the Christian church toward the truths of modern science."[28]

Winchell concluded by observing that "it is the chief business of the church to look after the spiritual well-being of the people and not to fit them to carry forward the complicated machinery of modern civilization," adding that "one of the boasted advanced steps of the nineteenth century is the separation of the church and state."[29]

President Folwell and the board of regents may have thought they were merely hiring a geologist when they extended a contract to Newton Winchell in 1872, but what they got was much more: a broadly informed intellectual and a persuasive advocate for the scientific endeavor. Scholars would cross him at their peril.

WINCHELL'S INTELLECTUAL CURIOSITY was all of a piece, not confined to a professional and public sphere. He and Lottie continued to share a complex intellectual home life as they had from the beginning of their marriage, despite his survey work, which kept him away for months each year. Lottie was actively involved as the editor of his voluminous outpourings of geological research—the annual reports to the legislature usually ran to hundreds of pages.[30] Indeed, the one annual report she might not have edited, when she spent the fall in Michigan in 1874, is conspicuous for its spelling and punctuation errors.

In 1877, Lottie took the children, now ages three through ten, to visit her aging parents on the farm at Galesburg, Michigan. Hortie and Ima were school-age, and apparently Lottie tutored them that fall, since the family remained in Michigan until mid-November. Meanwhile, Newton rented out rooms to students at the State Street house. That fall, he rented to Benedict Juni, who had grown up on a farm near Morton in south-central Minnesota.

Juni had been "a practical public school teacher."[31] Newton noted this in a letter to Lottie, knowing that she, like he, could appreciate all that the phrase would encompass (wide-ranging knowledge, ability to control a classroom, a good work ethic) and Newton hired him to serve as the survey naturalist the following year. But Juni had even broader experience: his childhood had been marked in 1862 when, at age ten, he had been taken captive by the Dakota during the Dakota War and had spent seven weeks in their company before being released.[32]

Juni, Swiss-born, spoke German, in addition to English and Dakotan. Newton, who had begun to learn German as an undergrad in Michigan and appreciated its usefulness for a scientist, saw an opportunity. He wrote Lottie in Galesburg to convince her of a plan. "[Let's] employ his knowledge of German in familiarizing us & the children with conversational German," he proposed. "Can we not set aside one hour per day for an exercise of that kind—having the children participate? Then hire a German girl and transform our whole domestic intercourse into German as fast as we can."[33] Newton suggested giving Juni free room and board in exchange for instruction. "Let the children stay home from school and study German at home, with your assistance and mine. I think that in a year we could make something out of German," Newton coaxed Lottie. The children were at an excellent age to pick up a second language. Hortie was twelve; Ima, ten; Avis, five; and Alex, three. "Let us try to accomplish something besides daily routine living."[34]

ROCKS OF FIRE

The North Shore

1878–79

Lake Superior spreads in a shimmering blue sheet, lapping at the foot of Duluth. In June, the massive lake still chills the air at the water's edge, though the solstice sun rises high in the heavens.

The cool breezes off the water are a relief to the team of scientists dressed for fieldwork in jackets, vests, and neckties. Newton Winchell has returned to the largest freshwater lake in the world, and in 1878 for the first time as Minnesota state geologist, he will be sailing on big water. He had made at least one excursion to Superior as a geologist in Michigan, but now he heads a team that will eye the rugged rocks of the North Shore and attempt to explain the geology, to assess the area's natural resources, and perhaps offer suggestions on how best to utilize them.

Winchell scuffles about, organizing his equipment, bantering with Benedict Juni, directing the camp set up. Chester Creek runs burbling past, a good source of water for the crew.

The summer field season begins.

NEWTON WINCHELL; CHRISTOPHER HALL, a science professor at the university; Benedict Juni, the survey's naturalist; and John Mallmann, a local Duluth resident who served as guide, had been in camp near Chester Creek on Superior's shore since June 19. Upon arrival, they had procured a boat, to sail or row, and taken it to this stretch of beach to set up camp. The next day, Mallmann went into town to purchase supplies—chisels, picks, four-quart pails, a fry pan, and the like—and the others commenced geologizing.

On foot, Winchell followed Chester Creek up the hill from the lake on a road running on the east side of the creek. The bugs were horrendous, but he was becoming inured to hordes. He noted that as the stream descended to the lake, the rock was very hard, dark colored, and fine grained. He called it "quartzite or petrosilex,"—today, geologists call it "basalt"—and observed it was the same kind of rock that appeared in dikes in the prominent rock formations closer to Duluth's harbor. This was his first extensive look at the dominant rocks of Minnesota's North Shore. Clearly, the rocks were born of heat and pressure, deep in the earth's core, the product of volcanoes spewing lava, which would later harden.

But he soon discovered that heat and pressure were not the sole agents of Lake Superior's geology. Straying away from Chester Creek, he wandered into the hills above Duluth and was surprised and disappointed to find very little bedrock. What he encountered instead was mostly a red hardpan glacial till. He assumed it was a legacy of glaciation, and he was correct; we now know that the reddish soil was deposited when Glacial Lake Duluth formed after the most recent glacier retreated.[1]

Winchell collected a rock the locals referred to as "Rice Point Granite" that had been quarried at a location between Rice Point and Park Point, a narrow spit of sand that begins at present-day Canal Park and extends like a finger toward Wisconsin. "This we call No. 1," he scribbled in his field book. It was hard gray rock; "The mass appears to be Labradorite," he noted, "But it also has magnetite."

The specimen was, in fact, gabbro, a dark igneous rock that cooled slowly deep in the earth and after uplift, was exposed by erosion. Gabbro, Winchell would discover, was one of the prominent components of what would later be called "the Duluth Complex," an association of rocks comprising most of northeastern Minnesota.

A visitor arrived bringing mail that had been carried north from Minneapolis on the train. Winchell received a letter from Lottie, which he devoured. For the rest of the summer, mail delivery would be irregular. Duluth had a post office, and there were opportunities to get mail at Beaver Bay, Grand Marais, and Grand Portage, the only communities on the shore. Winchell instructed Duluth's postmaster to give any letters addressed to the survey party at Beaver Bay to the carrier moving up and down the shore via boat. He could locate the party even if they weren't in town. This mail carrier, likely Chief Beargrease,[2] father to John Beargrease, for whom the annual dogsled race along the North Shore was later named, was based at Grand Portage and carried supplies and letters to the few residents of the North Shore.[3] It took a week to travel from Grand Portage to Duluth and back.

The party would watch hungrily for the carrier and eagerly consume the news he brought of life beyond the rocks and water of Lake Superior. They entertained him if he arrived at mealtime, pumping him for information. "I wish you would send me the *N. Y. Tribune* weekly," Winchell instructed Lottie. "We want the news."[4]

He also longed for letters. A common theme to the summer field season, whether Newton was in southern Minnesota or the wilds of the Arrowhead, was that Lottie didn't write enough, never as frequently as he did. "Why don't you write to me?" he asked the mother of four running a large household without help. "I think I have written half a dozen times to you since I left home & it is now more than 2 ½ weeks since but I have heard nothing from you."[5]

In 1878, the children were also old enough to write. "Can't Avis [age six] write me a letter? Ima's [age eleven] was too short," he wrote to Lottie.[6] When the girls then produced notes, Newton enthused, "Tell Ima I was very much surprised to have a letter from her & Avis alone. They are nice little writers and I am glad Avis can print so well."[7] A very involved father, especially for the times, he missed his children keenly.

The survey's first camp at Chester Creek was not within Duluth's city limits. The team, however, had neighbors nearby, and Winchell knew them. Judge Ozora Stearns, his wife, Sarah, and their three children lived on the outskirts of town in an elegant house at what is now Eleventh Avenue East and London Road. The judge was a prominent lawyer and circuit judge for the eleventh district; Sarah was an ardent advocate of women's suffrage. She had attended the Minneapolis rally in 1876 when Lottie was up for election to the school board, and the women may have formed a friendship then. Also, Ozora was a graduate of the University of Michigan's law school, and Winchell was likely to have been acquainted with him from the 1850s.

Another connection with the Stearns may have made Winchell smile. In 1858, when he had been an undergraduate at the University of Michigan, a group of women had demanded entry into the all-male school. They had been refused, though they made enough of a ruckus that Winchell mentioned it in his journal.[8] Sarah Stearns had been one of those women.

Mrs. Stearns issued invitations to the survey crew. She asked everyone to Sunday tea their first Sunday in Duluth, fortunate timing for the men, who still looked respectable. Their skin had not yet reddened, and their beards had not turned woolly from neglect. The Stearns children visited the camp to inspect the tents and the general milieu. After renewing ties with the Stearns, Newton invited Lottie and the children to come up on the train for a summer stay.

"I am sure you better visit Mrs. Stearns sometime this summer. It is a pleasant place (Duluth) to spend a few weeks. The air is cool and the lake is very pleasant. Large steamers are passing frequently, and from their house there is a fine view of the lake. The children also would enjoy such a visit," Newton noted, but then, possibly considering three lively Stearns children with four noisy Winchells, he added, "but perhaps they could not all come."[9] Later, by letter, he continued to urge Lottie to take the train to Duluth. If she came in mid-August, he would arrange to be in town. "We may be along there about Aug. 20th," he wrote. "We have something of a reorganization for an inland trip up the British Boundary [the Canadian border] toward Lake of the Woods."[10]

The crew enjoyed their first days in the field. The weather cooperated with cool nights but little rain. The scenery was stunning: sunlight glittering off the blue water, and white-sided steamboats chugging to and from the harbor. The many moods of the big lake beguiled them—the placid mirror surface transforming to pounding surf in a matter of minutes with a shift of the wind.

The men also enjoyed the unexpected skill of their designated cook, Benedict Juni. He prepared ham and bread for breakfast, and fish, potatoes, and coffee for dinner. He baked white, yeast-leavened bread over the fire on the beach, and after the men trolled for "speckled trout" with a lure of a red feather and a piece of tin,[11] he prepared the catch for a special dinner. One fish fed all four.

Winchell and his men planned on finishing their work in the Duluth area in a week's time. After two days in the field, he blithely wrote Lottie that "it will take all of two more here."[12] As he worked farther afield, however, he encountered more problematic formations. He found metamorphosed shales—mud and clay that had been transformed by high heat and pressure, interbedded with basalt that had been lava, oozing from volcanoes. There were sandstone and quartzite (Winchell's term for quartz veins) and numerous dikes of basalt, large slabs of rock that jutted upward through the subhorizontal bedrock of lava flows that dipped gently toward the lake. Geologists would later understand the shales to have formed during quiet intervals between volcanic eruptions.

Winchell collected more and more samples, all of igneous origin. He added specimens pockmarked with greenish inclusions. He thought it might be epidote, common in similar rocks on Michigan's Keweenaw Peninsula. He picked up rocks with reddish crystals, with calcite, with fine crystals, with coarse crystals. He looked in quarries, which exposed rock features otherwise hidden. He poked about the rock cuts near the depot.

Three days later, he wrote Lottie, "We shall not get away from here for a few days yet. We are getting puzzled by the complexity of the geology & must go carefully over some of our work."[13]

The team finally pulled up stakes and moved camp. They traveled by boat following the shore, and not by wagon. Only the roughest of trails connected Duluth with the scanty settlements on the North Shore. It would be a sailing summer.

Winchell now recalled how the wind could impede work. Leaving Chester Creek, the crew headed northeast for Lester River, to progress only two miles from the original camp, forced to row against a high wind for an hour and a half.[14] Lottie fretted about the situation, no doubt having heard stories about his youthful misadventures on the Michigan survey—the capsizing, the sudden landings, the loss of gear. She had written about her dread of "that lake." Superior's ferocity was already legendary. Winchell assured her they were "comfortably fitted out. Have a good boat (21 ft. long) and shall not endanger ourselves. We don't want any mishaps any more than you do & intend there shall be none . . . [we] do not expect to work on the lake when there is any danger."[15] With that dubious assurance, Winchell and team sailed northeast toward the settlement at Beaver Bay.

It was neither copper nor iron that had triggered the legislature order, however, but an 1865–66 gold rush that had drawn speculators to the rocks surrounding Lake Vermilion. Ironically, geologist Norwood, who had leaked the rumor of gold, had downplayed the "immense bodies of magnetite and hematic iron ore,"[16] and these deposits continued to be ignored throughout the 1870s.

The Lake Vermilion gold rush did not pan out. By 1867, the prospectors were gone, but the initial frenzy had created a road leading from Duluth to Lake Vermilion, the first inroad into the vast deposits of iron ore on the Vermilion and Mesabi Ranges, awaiting exploitation only six years hence. (That road, which followed an old Native American trail, today has three names: the Vermilion Trail, St. Louis County Highway 4, and Governor Rudy Perpich Memorial Drive.)

When Winchell set out in 1878, there was no organized industrial mining in northern Minnesota.[17] However, there had been scattered mining since the gold rush: attempts at copper mining on the French River and also on the Cascade River, and silver mining in similar rock in "British territory" (northwestern Ontario). There was considerable activity on Pigeon Point, at the international boundary. There, speculators had encountered minerals like calcite, barite, and amethyst, as well as chalcopyrite—the prominent copper-bearing ore—and reportedly silver.

On the Michigan side of Lake Superior, however, a boom in copper mining had been under way for thirty years. Communities like Houghton, with the big Quincy mine, and Calumet, with the rich Calumet and Hecla mine, headed by Alexander Agassiz, son of Louis Agassiz, prospered from copper money. Scores of active mines on the Keweenaw Peninsula pulled millions of dollars of copper ore from rock formations identical to those in northern Minnesota. In the boom's early decades, these mines had contained masses of pure copper, sometimes as much as hundreds of tons. The exuberant copper mines of the Keweenaw were no doubt on the minds of would-be entrepreneurs in 1878 as Winchell and his crew, notebooks in hand, tromped about the unbroken land of Minnesota's Arrowhead.

Minnesota's early mining attempts and areas of active exploration were stops on the way as Winchell worked the shore, taking samples at each, boxing them up, and sending them home to the university for analysis. The legislature wanted information on all possible commercial mining opportunities within the state.

The survey sailed into Agate Bay, future site of the town of Two Harbors, and looked at the rocks. There they encountered a party of tourists, the superintendent of Hennepin County schools and his family on a boating day trip from Duluth. The chance meeting became an opportunity for the geologists to send off mail, which they did. Agate Bay was also a great place to collect, yes, agates, and Winchell pocketed some for four-year-old Alex back home.

The survey sailed on. Shortly before reaching a river spilling spectacularly over a series of ledges, Winchell noted in his field notes that the shoreline was formed by heavy traprock. *Trap* is a general term for what Winchell would later pinpoint as "basalt." They had followed this trap almost continuously since leaving the Encampment River, and as the crew approached the Gooseberry River, Winchell detected small grains of copper embedded in the trap.

He estimated the falls at about one hundred feet, descending in a series of cascades. The several streams rustled as they spilled frothy white from ledge to ledge. The river came to rest on a gravel beach and spread out to meet Lake Superior. This area would later be protected as Gooseberry Falls State Park.

Winchell didn't comment on its beauty. He did note that someone in the party saw a woodland caribou at the mouth of the river.[18]

A few days later, at the mouth of the Split Rock River, Winchell lounged in camp and wrote more letters. The mail carrier was expected, heading toward Duluth the next day. It was Sunday, the day of rest, and Winchell's crew did not work.

Hot July weather swaddled the camp at Split Rock, and though a storm had threatened, wind, and not rain, had buffeted it. At Split Rock, Winchell described "a high masonic point" that appeared to be granite from a distance—it was whitish and protruded through the dark traprock—but that he called "felspar" (feldspar) and recognized as having been singled out by a previous North Shore geologist, J. G. Norwood.

Norwood described the feldspar-rich rock as a dike protruding through the traprock, which would make the feldspar younger than the trap. Winchell disagreed. He recognized it as older.[19] The feldspar had been formed deep in the earth and had been carried to the surface by rising magma and thrust between the trap. More resistant to weathering, the light-colored feldspar formed a prominent feature on the coast. Thirty years later, it would be deemed the perfect site for a lighthouse.[20]

The crew arrived at Beaver Bay on July 17, a full month after first arriving in Duluth. There the men got more mail and supplies, because Beaver Bay was a thriving community, the only one between Duluth and Grand Marais. A German family of five brothers, the Wielands, had settled the area in 1856, farming and logging. They operated a water-powered sawmill, ran a general store and a post office, and sailed a two-masted schooner, the *Charley*.[21]

The Wielands often employed Ojibwe affiliated with the Grand Portage Band who summered at Beaver Bay. Local business was conducted in a mix of Ojibwe, German, and English, the Ojibwe learning German, and the townspeople, Ojibwe.[22]

Northeast of Beaver Bay, Winchell looked for the dominant profile of Palisade Head rising from the water, a well-known formation of the shoreline. He collected samples from the base of the bluff, wielding his hammer and chisel balanced in the rocking boat; and he collected from the summit, where he described the stony surface as "hard reddish brown with translucent rectangular crystals, fine-grained generally."[23] Today this is known as the Palisade Rhyolite. He observed that the palisade bluff was composed of columns of rock. These columns broke apart lengthwise as ice and frost weakened their structure, and then, assailed by Lake Superior, they dropped vertically to the foot of the bluff. On this first visit, Winchell could see a column in the process of detaching, leaning against the bluff and somewhat intact.[24]

The geological survey beheld spectacular natural wonders: the several spills at Gooseberry Falls, the vertical drop on the Baptism River, the massive Palisade Cliff. But the wonder that blew Winchell away was the Temperance River. Dryly noting that the river, unlike all other North Shore rivers, "has no beach or spit of gravel" at its mouth, he declined to acknowledge the little joke of its name. (The river has no bar.) Winchell described the rushing river's cut through massive traprock, the several roaring cascades as the water made its way to the lake, the potholes excavated over time that pocked the surrounding rock.

Perhaps it was the visual thrill of the Temperance's gorge that mesmerized him. Standing on its edge, one feels tempted to try to leap the gap. "The gorge & falls & the potholes are . . . the most remarkable combination of picturesque erosion, and inaccessible wild scenery that I ever saw," he wrote. "The gorge is so narrow it can be stepped across, the only danger being to secure footing on the other side, for a failure to *stay* on the other side would precipitate a man down a gorge from 50 to 100 feet into a foaming river." Wishing to capture the thrill, he added, "Several fine photographs should be taken."[25]

The survey sailed on. In late July, to escape high winds, the party was driven onto a beach northeast of what is now Lutsen. The experience took a surreal turn when the midafternoon air darkened while they were beached and the sun was eclipsed for about half an hour. The wind did not abate, and the men remained there for the night, as breakers pounded the shore. They pitched their tents on the bluff above the beach and whiled away the forced encampment by collecting myriad thomsonite stones scattered about, gemstones that Winchell described as "pink and white radiating masses." To commemorate the freakish day, Winchell named the spot "Eclipse Beach," but the name did not stick.[26]

At Grand Marais, the geologists sailed into its perfect natural harbor, protected from Superior's pounding waves by solid traprock. The red pebble beach was kind to their feet and kind to their boat. There they packed twenty boxes of rock specimens and sent them to Duluth by boat and then on to Minneapolis. They had spent time on Good Harbor Bay, five miles southwest of the harbor, collecting additional "beautiful thomsonites thickly studding . . . a bed of apparently igneous rock."[27]

In 1878, Grand Marais consisted of two log houses: one on either side of the bay, and a rough cabin down by the water. Two bearded, white-haired brothers, Henry and Thomas Mayhew, lived on the north side of the bay, and a third man, Samuel Howenstein, lived on the south with his Ojibwe wife.[28] Henry Mayhew and Howenstein had come to the area in 1871 to launch a commercial fishing operation.

The Mayhews ran a trading post out of their cabin, and they had recently built a post on Rove Lake on the Canadian border as well. In the winter, they would travel over a rough trail to reach the post, hauling supplies in and furs out, via a sled pulled by a big dog and a billy goat.[29]

The third log building, loosely joined, functioned as a fish house. The Mayhews loaned this shed to the survey team. It had three rooms and a woodstove for cooking meals. The men slept there when in the area and used the packing and salting room as a "parlour" to entertain guests. Never intended as a dwelling, the place let in the wind and the rain but still served as a base of operation.[30]

The Mayhews proved immensely helpful to Winchell and his crew. They knew the area well and advised the men on conditions, outcrops, and other information useful to the expedition.

The party continued northeast along the shore to Grand Portage Bay. They camped on the north end of an island that lay in the bay, and Winchell's hammer was busy chiseling off samples of hematite, calcite, slate, and basalt—a plethora of igneous and metamorphic rocks and minerals. Numerous dikes, some many feet wide, others narrow and measured in inches, crisscrossed the bedrock.

Winchell traveled to speculative mines in the area to examine them. He took a boat out to the Susie Islands, just off Pigeon Point, to collect specimens. He went up the Pigeon River and sampled each waterfall ledge. He accumulated more than fifty samples in all and added them to the 250 samples already collected, about forty-five boxes for the field season.[31]

In mid-August, Winchell, Hall, Juni, and Mallmann parted ways. Winchell caught the boat for Duluth to meet his family as planned.[32] Later that month, he returned alone to Grand Portage to commence the remainder of the field season.

BEFORE THE START OF THE 1878 field season, the state legislature had ordered the survey to transfer its efforts to the northern part of the state. Winchell had not yet seen Minnesota's Arrowhead region, those counties bordering Lake Superior's north shore. However, speculators were already eying the area. As early as 1850, geologist J. G. Norwood identified iron-bearing rocks on Gunflint Lake.[33] Native Americans, whom Winchell called "the ancients," had mined pure copper veins ("native," or metallic copper) on Isle Royale since prehistoric times, fashioning fishhooks and spear tips and other useful items. Sharing Lake Superior's shoreline, Michigan's Upper Peninsula had rock formations like those of Minnesota and was in that decade experiencing a boom in iron ore and copper mining. People assumed that valuable ores—moneymakers—were hidden in Minnesota's bedrock.

10

PORTAGES

The Boundary Waters and Iron Ranges

September–November 1878

THE SURVEY TEAM WAS NOW RECONFIGURED into two independent parties. Christopher Hall led a crew to ascend some of the rivers that entered Lake Superior from the Minnesota shore. Beginning near Grand Portage with an Ojibwe guide, he worked his way toward Duluth. Most of Hall's focus was on elucidating the geology of the exposed bedrock in the streams, but he did visit one abandoned copper mine on the Rosebud River.[1] Later, near the end of October, this team spent time on the Devil Track River and Devil Track Lake. There, they tried out a custom-made collapsible canvas canoe, designed to fold and carry into any inaccessible spot. Hall found that the canvas canoe was slower and harder to row than a birch-bark one; still, it was light, strong, and well made and might prove useful under the rugged, wilderness terrain of the northern part of the state. Hall's party called it quits on October 25 in the midst of a howling snowstorm.[2]

While Hall was employed on the shore, Winchell prepared to trek along the international border lakes west to Basswood Lake, and from there, into the interior of the Arrowhead region. He wanted to sample the Vermilion and Mesabi iron-bearing belts to confirm what previous geologists had found.

He assembled his team at Grand Portage, hiring two men to act as guides. He did not identify the guides by name, but they undoubtedly were Ojibwe from the community, intimately familiar with the portages networking through the boundary waters. The three would travel by birch-bark canoe. It was September, with freshening cold fronts breezing in from the northwest, birch leaves yellowing, and lakes cooling. Soon, cleansing dips in the water would become less pleasant and more a miserable necessity.

The survey work Winchell had set for himself was physically demanding. The Pigeon River, delineating the international border, presented the first challenge.[3] The river descends about 660 feet to Lake Superior in a series of rapids and waterfalls, culminating in Pigeon Falls, which, at about 90 feet, is the highest falls in Minnesota. This waterway had been known to Native Americans for centuries, and later, to French-Canadian voyageurs, who paddled canoes heavily laden with furs, bound for Montreal and Europe. The last section of the rough, tumbling river could not be navigated by canoe, and Indians and voyageurs alike used the eight-and-one-half-mile Grand Portage that circumvented the roughest length of the river.

Winchell and his guides took the Grand Portage in early September 1878, beginning at Grand Portage Bay of Lake Superior. One or perhaps both guides carried the canoe. On day one, the cargo was as light as the party would haul, since the boss had not yet begun chipping off samples of rock with his hammer to take home and analyze. Later, the men would lug big, heavy canvas sacks of rock over portages in addition to their gear.

Winchell matter-of-factly described the trail as rising 537 feet above Lake Superior in the first half mile: "the ascent being gradual all the way, and nearly in a right line."[4] With his geological eye, he observed that a glacier valley, partially filled with till, determined the location of the ancient trail. At the summit, which he assessed as five miles in from the lake, the hills all around, once wooded, were charred and treeless from an accidentally set fire. This summit, he noted, was a frequent resting place for voyageurs.[5]

Beyond the summit, the trail descends to the Pigeon River, and the guides deemed the water levels adequate for paddling. The party dropped in and pushed upstream against the current to the next portage, around Partridge Falls. Then, with birches golden and frost in the air, they continued paddling and portaging through South Fowl Lake and North Fowl Lake, and on to Rove Lake to reach the Mayhews' trading post in mid-September. The brothers from Grand Marais had set up shop on Rove Lake only recently, intending it as a winter trading post.[6]

The party repacked and replenished supplies. Winchell wrote Lottie his only letter from the trip. "We are all well," he assured her. "Have had good health & good success all the way." He was concerned about eleven-year-old Hortie's health: "If he remembers what I told him about keeping in the house he will come out all right, but I don't think he will remember." He was concerned about Lottie overtaxing herself: "Get a good girl" as household help, he admonished her. Benedict Juni was boarding with the family that fall, and Winchell assumed the German lessons continued. "Remember me to Ima & Avis & Alex & Hortie" he wrote, adding, "I do not know anything about when we shall be out of the woods." He was then incommunicado for the next five weeks.[7]

On a map, the Canada–United States border appears to be a straightforward shot comprised of lakes and portages. But in tromping along portage trails, with their feet squishing through soggy low areas, clambering over downed trees, and stumbling along rocks embedded in the earth, Winchell and his guides experienced the full extent of its rugged, challenging terrain. Portage trails alternated with paddles on the long, narrow lakes, often with headwinds that whipped up whitecaps slapping the canoe. To Lottie, Winchell described the international boundary as "a crooked Indian trail between this lake [Rove] and the next [probably Rose Lake]."

The border lakes lie in what is now called the Rove Formation, a sheetlike slate interrupted by layers of a much harder type of rock, diabase, formed by intrusive sills of molten rock into the slate. This formed a geological version of a rocky layer cake. The cake had then been tilted, exposing a series of parallel east–west ridges: slate, diabase, slate, diabase. The lakes were positioned between the ridges, which were held up by the hard diabase sills. This made travel in an east–west direction fairly easy, but traveling south would involve climbing rugged hills. The survey would continue trending westward until it came to Basswood Lake.

Temporarily leaving Rove Lake and the international boundary, Winchell and crew portaged south into Daniels Lake and on to Bearskin, then east in sequence through some lakes that apparently were not named to Pine, McFarland, and John Lakes. Winchell was interested in examining mining shafts near some of these lakes. One had yielded pieces of native silver,[8] but none had produced copper. Winchell noted that no prehistoric implements, like those found in copper mining pits on Isle Royale, had been found.

One prospector, John McFarland, owned several veins and a site thought to be one of ancient mining on or near the lake named after him, but these had not yielded quick profits, and he had turned to farming and fur trapping to make a living.[9]

This little side jaunt brought the party through a lovely landscape of mature white pines with graceful arcing branches, dark spruce collected in low areas, spires to the sky, and massive rock ledges, perfect for campsites.

Having checked out these scattered mining attempts, the party then returned to the border lakes and continued west. Between South and North Lakes, Winchell noted the continental divide. This would later be called "Height of Land" portage. The map he was using had incorrectly plotted the watershed that separated waters flowing into Lake Superior from those into Lake of the Woods.

The south shore of North Lake had a sandy beach with chips of black flint and reddish jasper. Winchell recognized this as the rock early explorers and voyageurs had used to ignite the gunpowder in their guns.[10] A series of portages then brought the party into Gunflint Lake, a long, beautiful body of water with dark, high cliffs and abundant white pine. In a mere eight years, Henry Mayhew would discover iron ore near this lake, and open a mine two years later. The seemingly untouched boundary waters were poised on the brink of development.

The survey continued west along the international border through big Lake Saganaga. Winchell commented on the lake's many islands (which *saganaga* means; Winchell added that it seemed to be the plural of *saginaw*) and the jack and Norway pines. Paddling west out of Sag, the survey took Oak Portage, so named because of the bur oak growing there—the first oaks seen since the party left Grand Portage. The portage led into a little lake surrounded by hills. Winchell climbed one of them. At the top, he was surprised to see the entire landscape barren, rocky, or burned for as far as he could see.[11] Down the hill from this moonscape, the men paddled into Ottertrack Lake, narrow and five miles long. Ottertrack led to Knife, another narrow, complicated lake with islands, bays, and a large south arm. Knife was marked by slate outcropping that was hard and sharp, hence the name.

By this point in the journey, Winchell had seen more of the boundary waters than most Minnesotans ever would, and he wrote in an annual report that "the size of these lakes is often a surprise to the traveler. They expand unexpectedly, where the prospect is entirely shut off. They are shaped by geological features. . . . Sometimes a narrow opening in one ridge allows the lake to spread. Then it enters another narrow valley, and runs in it visibly a couple of miles, when it may jog back again into the former valley by another opening. Only one acquainted [with the area] could follow the boundary line canoe-route."[12]

At Basswood Lake, west of Knife, the party needed to angle south, away from the border and toward the iron belt at Lake Vermilion. Basswood had an abandoned Hudson's Bay Company fur-trading post located about two miles from its eastern end on the Canadian side of the international border. There was also an abandoned post opposite on U.S. soil. Winchell surmised that frequent forest fires had destroyed the fur-bearing animals and thus brought an end to commerce there.[13]

The survey followed Pipestone Bay of Basswood to a lake he called "Kawasachong," Ojibwe for "mist or foam" and what today is known by its English translation, "Fall."[14] Kawasachong was notable for its prominence in a canoe route that led from this lake to Lake Superior at Beaver Bay.[15] Had Winchell and crew wanted a quick route to the big lake, this would have been it. But instead, they paddled the length of Kawasachong and took a short portage into Long Lake, known today as Shagawa, future site of Ely.

Winchell and companions ventured south to Vermilion Lake because it had been the location of the 1865 gold rush and because other mining attempts had occurred in the locale. Earlier geologists had also established the presence of sizable deposits of iron ore, both on the Vermilion Range and on the Mesabi Range to the east, and Winchell needed to further refine the knowledge of these.

Lake Vermilion offered many attractions to the geologist. Winchell sampled exposed rock at the shoreline and checked out portages leading south, west, and north out of the lake. He visited the Government Station on the southern portion of the lake, "where the Indians are taught to do some farming," he noted.[16] The New York Mining

Company had pursued gold in the vicinity of the station in the "rush" of 1865.

The broken dreams of several mining companies were visible within a few miles of Government Station. Heavy equipment marked the demise of the Nobles' Mining Company: two large reverberatory furnaces and five stamp mills, made in Chicago and hauled up the primitive Vermilion Trail from Duluth.[17]

The survey crew reached Lake Vermilion in October. Fall was advancing, and furious waves raced across its large expanse of open water under the blow of northwest winds. Nights were getting longer and colder. Branches were bare on the aspen and birch. The dark spires of conifers seemed sparer, starker.

Earlier geologists had located the Vermilion iron belt on the southeast end of the lake. Winchell decided it ran northeasterly toward Gunflint Lake and deemed it at the point where he was to be about twenty-five feet thick. The bedding reminded him of what he had seen at Marquette, Michigan, which in 1878 was the site of tremendous iron ore mining. He commented on the "beautiful variations between jasper and hematite" in reds and terra-cottas.

The party left Lake Vermilion's big water behind, doubtless with relief, and paddled down Pike River, taking a series of portages and running a few rapids with the heavily laden canoe. They portaged into the Embarrass River, so named by voyageurs for the many obstacles it presented to canoeists. They traveled southward, crossing the Vermilion Trail. The Embarrass took them into the Mesabi Range, where Winchell examined shallow pits and trenches that had been dug by prospectors. His compass needle spun round and round, essentially useless in the presence of magnetic rock.[18]

Like others before him, Winchell found much iron ore. He observed, "The ore resembles those of Scandinavia and Russia, as well as the magnetic ores of northern Michigan. . . . For making steel, these ores excel. . . . It is highly probable that these iron deposits will not lie long undeveloped."[19]

After days spent on the Mesabi Range, Winchell and crew continued their paddle on the Embarrass River. Just before its confluence with the St. Louis, the Embarrass ran past countryside that appeared to have been ravaged by frequent fire, destroying the pine and encouraging fringe growth of aspen everywhere.[20]

The party then portaged into the St. Louis River, and from the St. Louis near what is now Floodwood, they paddled up the East Savannah River to the Savannah Portage (used, like the Grand Portage, by Native Americans for hundreds of years, and by French fur traders in the 1700s and 1800s) to Big Sandy Lake and the Mississippi River. This portage is both historic and notorious for its demands on the traveler. When Winchell slogged through it, he noted that it was not the old portage trail but a new one, seven miles long, passing through a cranberry bog, and then up and down over a kame, a sand and gravel hill left by the melting glacier. Aspen, white pine, and tamaracks flanked the path, which despite its recent origin, was well trod. The seven-mile trail terminated at Prairie Lake, which flows into the Prairie River. When water levels were high, this river could be paddled, but that October, it needed to be portaged. After the preceding six weeks—the border trail, the huge lakes with dangerous waves, the long portages with rock-filled packs—there was yet another sixteen miles of walking to reach Big Sandy Lake and the Mississippi River, which would be their conduit home.[21] At least the mosquitoes, horrendous in summer, had been killed off by the frost that now rimed the landscape.

Not given to excess emotion, Winchell laconically took the rigors in stride. "We are well and have had no bad luck, tho the trip has been tedious and rough," Winchell wrote Lottie from Aitkin, after completing the final leg of the trek through the Arrowhead region.[22]

Perhaps in this autumn of 1878, when Winchell tromped the twisted undergrowth of bogs, scaled the impediments of burnt and wind-downed logs, braced headwinds over long expanses of border lakes, and labored every day with the minutiae of meticulous field notes, his reputation among the Ojibwe took root. His physical and mental toughness and stamina would become legendary and be remembered at his memorial service.

"The Indians admired him for his physical strength and endurance," an acquaintance recalled. "They said he would travel all day and lie on the ground, with his feet to the fire and a stone for a pillow, and sleep all night, while other white men wanted a soft bed made of fir boughs. . . . Asked to name a man with a perfect physical constitution, I should have named him before any other man."[23]

11

BACK TO THE NORTH SHORE

Summer 1879

WHAT AN ABUNDANCE OF RICHES! "The State of Minnesota possesses the widest range of lithological features,"[1] Winchell exulted in the opening of his eighth annual report to the legislature. The state showed nearly every possible example of rock type and change in rock type through geologic forces seen in the United States. Ancient sea floors? Southeastern Minnesota had them. Glaciated landscape? There, in western Minnesota. Molten rock from the earth's interior? Superior's North Shore. Glacial lakebed? The Red River valley.

Economically, Minnesota's geology was also promising. Scientifically, much of the land was unexplored territory in a fairly new field that was beginning to flex its explanatory muscles. Winchell promised that when complete, the survey would deliver to the state "the full and searching scrutiny that modern science can give."[2]

To grapple with Minnesota's geological complexity, in 1879 Winchell expanded his staff. He hired Warren Upham of New Hampshire's geological survey and assigned to him the counties in western and central Minnesota that had been buried under a thick layer of glacial till.

Upham, twenty-nine, had studied under the highly regarded Charles Henry Hitchcock at Dartmouth. Thin, with intense eyes, Upham had many interests. Arriving in Minnesota an unmarried man, he nurtured a love of history as well as words, and would one day leave as a legacy a tome tracing the origins of place names in Minnesota. Upham was broadly trained and would make his mark in geology by elucidating Minnesota's glacial past. He would name the melted glacial remains of the Red River valley "Lake Agassiz,"[3] after Swiss-born geologist Louis Agassiz, an early proponent of an extensive ice sheet enveloping much of Europe.[4] Upham would also determine the margins of the most recent midcontinental ice sheet in North America. In his first year with the survey, Upham traveled 3,300 miles by horse and wagon in Minnesota in the course of the field season, collecting samples and observing landforms.

Upham's arrival indicated a turning point for the survey. Winchell turned over twenty-two counties, sixteen thousand square miles, to Upham's ministrations. He also directed university professor Christopher Hall to turn his attention to the "natural history" portion of the survey, which Hall did, aided by unpaid workers, who contributed greatly to the field study.[5]

Winchell reserved the geology of the Arrowhead region for himself. Before returning to the North Shore in summer 1879, he wrapped up phase one of the survey with a quick trip to southeastern Minnesota to revisit Goodhue County. "It's a large and important county," he explained to Lottie. "I try to do the work thoroughly."[6]

It was May, and it was raining. The roads were muddy, and his horse needed to be reshod to keep the animal from slipping. In his letters home, Newton wrote messages to each child: Hortie, age thirteen, was responsible for splitting wood; Ima, age twelve, was sent a dime for mending his socks. Alex, age four, suffered from measles. "Tell him I am sorry he is sick . . . [but] they are so light, generally this time of year."[7] Avis, age seven, was sent a book, *The Story of Avis*—really, a grown-up book meant for Lottie.

Popular in 1879, *The Story of Avis* related the tale of a woman who wanted to have a career. Avis of the novel falls in love but turns down marriage to pursue life as an artist. The Civil War intervenes, and Avis reverses her decision to marry, with disastrous results. Newton writes Lottie that "the book is too old for [seven-year-old] Avis to understand now, but Mama can read it. I guess you will enjoy it. There may be some things in it that illustrate your life and experience, especially the curtailment of your artists' [*sic*] career and the utter worthlessness of your husband, and so the denial of both your mind's aspiration and your heart's love . . . if you don't think [this], I am glad."[8]

AFTER WRAPPING UP THE WORK in southeastern Minnesota, Winchell regrouped in Minneapolis, and in June headed north after school was out. This time he took the teenaged Hortie with him, possibly to make Lottie's life easier at home. His camps for June were near Fond du Lac and the Dalles of the St. Louis River in and near present-day Jay Cooke State Park. This section of the river cut through slate and graywacke bedrock that had been uplifted and fractured, creating tumultuous rapids and dramatic boulders. Rocks seemed tossed about by enormous force, some jutting heavenward, cleaving the sky.[9] The camps were always near completed track of the Northern Pacific Railroad, so they had access to Duluth and, equally important, to fresh milk, butter, and eggs to supplement their camp diet.

A number of people joined the survey in June, including two university students, Den and Adin Brooks, who sometimes were companions to Hortie and sometimes accompanied Winchell to collection sites. Their father, Jabez Brooks, a professor of classics at the university, also dropped in to camp that month.

Hortie was free to explore the wild countryside and *could* have served as a reliable assistant but did not. Hortie also did not write to his mother. "I cannot get him to write notes even except by driving him to it, and I am pretty busy, except when I am pretty tired," Newton wrote in resignation to Lottie.[10]

Undergirding the scientific endeavor of the survey was a meshwork of social connections, which undoubtedly cast a more genteel aura over camp life than is generally seen. One social and scientific connection was Congregational minister Cassius Marcellus Terry, who accompanied Winchell to sampling sites each day, serving as a field assistant, despite being severely hampered by tuberculosis. Terry had a connection to Warren Upham: his wife, Emily, a wildflower water colorist, was the sister to Upham's Dartmouth geology professor, Charles Hitchcock. The Terrys had also been friends of General George Custer and his wife, Elizabeth; they assumed the care of Custer's favorite dog, Cardigan, after the general's untimely death at the Battle of the Little Bighorn. Winchell valued Terry's cheerful disposition and urbanity.[11]

Reverend Terry should not be seen as a geological dilettante, however. In 1880, he would undertake an overview analysis of Minnesota's hydrology, spending weeks in the field. Winchell would praise the work for its carefulness and scope and add, "his perseverance and industry were remarkable, and the amount he accomplished with his feeble health was a surprise."[12] In the field, one observer considered Terry a better naturalist than either of the professors working on the team.[13]

In July, Winchell and crew had worked their way upriver to Knife Falls, the future site of Cloquet. In his two months along the St. Louis River, he had decided that the "great basin of the lake [Superior] actually does extend up the St. Louis Valley as far as Fond du Lac."[14] He had also attempted to correlate the rocks seen at Grand Portage with those in the St. Louis River valley.

The rest of the summer would be spent on the North Shore, often on the water in the survey's little boat. Perhaps to quell Lottie's uneasiness about safety, and perhaps because much of the time would be spent confined to the boat, Hortie's happy-go-lucky field days were over. He returned to Minneapolis and to the feminine domain of 120 State Street.

N. H. Winchell shuffles through papers, his writing supplies, his field gear, his shirts and socks. Where did he put that map? In camp again at Chester Creek, his gear is strewn about as he searches for that one essential item.

The weather is stormy. A northeast wind, sweeping the full breadth of Lake Superior, sucks the walls of the canvas tent in and out, in and out. Waves pound the pebbly beach at the mouth of the creek, breaking white and frothy. Members of the survey crew shout instructions at each other over the roar of wind and water. Rain speckles the air.

"Has anyone seen my map?" Winchell bellows over the din. No one has. Disgusted with himself, he frowns, hands on hips.

Finally, he settles down, pulls out paper and inkwell, and writes Lottie. "I find I forgot a map at home. . . . I think I put it in my cupboard case in the hall. It is a map of that part of L. Sup. containing Isle Royale & the coasts to Thunder Bay. I want it. I [hope] you can get it to me by the time I need to use it. It is a Coast Survey Map. *Fold it back & forth like a fan . . . so that when shut all up it will be small enough to go into my pocket . . . you need not cut it nor mount it on cloth, but you better fold it, carefully and closely, & wrap in paper tightly and send by mail to me at Duluth."*

He peers out the tent, where Superior heaves and foments. The lake is much rougher than when Lottie had been here last year in August. Wouldn't she worry if she could see it now?

WINCHELL REVISITED MINNESOTA'S Arrowhead region in July and August 1879 for a second look at key points along Lake Superior's shore. This time he brought along a cumbersome camera and all its paraphernalia to photograph particular geological features.[15] Five years earlier, Winchell had observed the photographer who accompanied the Custer expedition to the Black Hills. Now he was the one in charge of the camera, the box of plate glass, the portable dark room, and the chemicals.

In 1879, photography was on the cusp of change. Dry plate negatives that could be exposed outdoors and developed later in a lab were new. Winchell employed the old technique to make a picture that Mathew Brady had used in photographing the Civil War: he coated a glass plate with a photo-sensitive solution, placed it in the camera, and opened the shutter. After exposure, he removed the plate from the camera, and while the chemicals were still wet, working in a tent that served as a darkroom, poured on developer, fixing the plate, and then washed it in a bath.

There was a learning curve to the process, and his first attempt ended badly. "Tried my photo apparatus yesterday for the first time, but before developing the negative I unfortunately spilled the developer into the Hypo bath ruining both," he wrote Lottie. However, he was not deterred. "I have more of one & shall dissolve more of the other, so I can go on again."[16]

AFTER WORKING THEIR WAY UP THE SHORE to Pigeon Point, Winchell and his crew sailed out to Isle Royale in the little survey boat. They circumnavigated the island, disembarking to hike to abandoned mining sites, and took samples. He chipped at rock from the coast and inland. He was interested in the red sandstone from Siskiwit Bay that had been quarried and was being sold in Detroit.[17] The rock formed a rough, jagged coast, and Winchell estimated the sandstone unit was three to four hundred feet thick. Certain features suggested to him that the rock had been changed by "heated regions below."

Remnants of copper mining were everywhere on Isle Royale. One poignant ghost town surrounded the abandoned Island Mine.[18] Its stamping mill—where copper-containing rock had been crushed to a fine grit—still stood on the beach. Inland, by the mines, Winchell found empty houses, stores, even a courthouse, for it had been declared the county seat of Isle Royale, Michigan. The place was deserted, although he found the equipment to be in good condition.[19]

At the Minong Mine on the island's north shore, Winchell collected evidence of prehistoric mining activity. This mine was rich with copper that was pure, not bonded to other materials. It had not needed to be smelted to remove impurities, and since copper is malleable, it could be easily worked. Winchell picked up pieces of "battered" copper that had been fashioned by people he called "the ancients." Their stone hammers—oval pieces of basalt—lay in the open pits.[20] He packed an entire barrel of stone hammers to ship to Minneapolis.[21]

From Isle Royale, the survey sailed fifteen miles north to a peninsula in Canada with a fabulously productive silver mine on Silver Islet.[22] Much of the silver was pure metallic silver with no need of smelting. Winchell examined the mine to see where the greatest deposits were and what kinds of rocks were associated with them. He traveled to the bottom of the mine, 750 feet below the level of Lake Superior, and found diorite, an igneous rock. Over three million dollars in silver would be extracted from that mine in the next fifteen years.[23]

WHILE WINCHELL AND MEN SAILED on big water, Christopher Hall and three others spent August on the North Shore near Grand Marais and inland at Devil Track Lake, collecting primarily plants and birds. Reverend Terry accompanied this party but was quite ill and unable to do much.[24] Hall's assistant, Thomas Roberts, a student at the University of Minnesota, contributed to the botanical collection that Benedict Juni had made in 1878. However, Roberts was more interested in the birds of the region, which Juni had not collected. He shot 125 specimens and prepared the skins for the Natural History Museum. He also noted that a purple martin house had been erected by the Mayhews in Grand Marais.[25] Hall praised the twenty-one-year-old's "untiring industry and patience, "his quickness and exactness in observation," and "the thoroughness with which he followed out the details."[26]

After two summers on the North Shore, Hall was appalled by some of the wanton destruction he witnessed, and he employed blunt language in his report to the legislature. Outsiders in search of wealth or pleasure were squandering valuable resources, brazenly flouting protection the state had already conferred on them.

"It should be the effort of those engaged in the work of the [survey] to urge the preservation of what nature has given us in the shape of game, fish, forests, and scenery, as well as to point out the places where precious metals may be found, where fuel lies deposited or where the finest building stones may be quarried," Hall wrote.[27] The first of his concerns was the brook trout, a native fish inhabiting the many clear streams running into Lake Superior along its North Shore.

Hall noted that those streams were in 1879 one of the most famous fishing grounds in North America. Anglers engaged boats and guides in Duluth or in south shore communities, fished in all seasons, day and night, visited multiple streams, often not eating their catch but "carelessly throwing it away," and allowed the fish to decay on the banks, polluting the stream. "It is true we have game laws, but it is a very difficult thing to have them enforced."[28] The northern residents whose livelihood relied on fish and game were indignant but unable to protect them.[29]

Winchell had deplored the actions of wildcat mining prospectors in northeastern Minnesota, and Hall echoed his warnings. Prospectors came into the area, looking for gold, silver, or other precious ores on property they didn't own. They set fires to woods by a riverbank where a vein of quartz or calcite sparkled, promising riches. The fire cleared the trees—wasting valuable timber and creating immense devastation—and exposed bare rock, which could then be scrutinized as a possible mine site. Both men thought the state of Minnesota needed to take action, perhaps jointly with Canada, to halt the plunder. Hall suggested hiring officers to "ferret out violations of law and bring offenders to punishment."[30]

In addition to decimation of the Arrowhead's woods and fish, Hall noted a primitive road had been cut along the shore of the lake, from Duluth to Grand Marais.[31] Corduroy bridges spanned the rivers.

Sled dogs could travel the road to carry mail when Lake Superior was frozen, but the road was too rough for wagons. However, Hall believed that the road would "eventually develop into a substantial highway, as soon as there is a demand for it."[32] This is now Minnesota Highway 61.

Hall commented on how treacherous it was to sail on Lake Superior. "Throughout its whole distance, from Pigeon Point to Duluth, [there is] no shelter where the larger boats could find refuge from storms. The whole extent of this coast line is rocky: it has but two or three natural harbors, and is exceedingly liable to storms and heavy winds, especially in the autumn."[33] Hall complimented Congress for appropriating ten thousand dollars to improve Grand Marais's natural harbor by dredging, so that bigger boats could use it in a storm.[34] Work on the harbor had already begun.

Hall had also made the trek to Carlton Peak, one of the highest points on the North Shore. At its summit, he said, "a landscape of surpassing beauty lies before the beholder."[35]

AFTER TWO FIELD SEASONS in northeastern Minnesota, Winchell had amassed almost eight hundred rock specimens to analyze. Numerous fossils taken from sedimentary rocks in southeastern Minnesota also awaited identification. Cataloging the fossils would be easy but time consuming. Geologists had been studying fossils for more than a century.

Making sense of the igneous and metamorphic rocks Winchell had collected in the Arrowhead would be more challenging. They were composed of minerals—crystals with a definite chemical formula—and identifying which minerals occurred together would tell Winchell something about the rocks' origins, as well as their relationship to North American geology as a whole.

To do this analysis required new technology. Cutting-edge research employed a microscope that had been modified to allow the stage of the scope—where the specimen is placed—to rotate.[36] Light polarized by a prism was passed through the specimen. The crystals bent the light in certain ways. Measuring the degree to which the mineral crystals in the rock specimen bent the polarized light rays gave researchers an idea of what minerals were present in the rock sample, for each mineral has unique characteristics under polarized light.

This technique was so new that microscopes designed to do this could be bought only in Europe,[37] but Winchell had the survey's microscope modified in New York to make the measurements. Reverend Terry mastered the art of making thin sections of rocks, three hundred of them, so thin that light could pass through them, and affixed them to microscope slides.

Winchell now needed to spend time in the lab, identifying, analyzing chemical compositions, and aligning rock strata, drawing orderly conclusions, and making methodical sense of the grandeur of Minnesota.

12

THE BOOM

1880–84

MINNEAPOLIS WAS BOOMING! Newcomers poured into the burgeoning town in search of jobs they surely would find. Leaving muddy imprints in the unpaved city streets, men tromped the board walks of Hennepin and Nicollet Avenues, bellied up to one of the well over two hundred bars on tap downtown, and prowled the East District's Main Street red-light district flourishing at the foot of the suspension bridge south of East Hennepin Avenue. The 1880 U.S. Census had counted 46,867 Minneapolis residents. Within three years, that population was deemed to have more than doubled to 94,337.[1]

The city's economic expansion was sparked by an actual explosion on May 2, 1878, when the Washburn A Mill, the pride and joy of the flour-milling Washburn Crosby Company, blew up with a force that knocked pedestrians off their feet several blocks away and shook houses on their foundations all over town. Residents thought an earthquake had struck.[2] The catastrophe, which killed eighteen men and remains Minneapolis's greatest industrial disaster, destroyed a third of the city's milling capacity in the initial explosion and the fires that ensued.

But as is so often true, disaster turned into an impetus for change. New milling technology that greatly improved productivity was being developed in Europe. It employed steel rollers, not millstones. Indeed, several years before, rivals Charles Pillsbury and C. C. Washburn (owner of the flour company that would become General Mills) had jointly sent an agent to Hungary to check out the new rollers.[3] After the explosion, the investment seemed worth the gamble.

As Washburn rebuilt, the Pillsburys added a massive new mill themselves, the Pillsbury A, on the east side of the falls. This mega-mill—bigger than the new Washburn mill—featured one hundred steel rollers, a passenger elevator conveying workers to the upper levels, and electric lights, two bulbs on each floor, making it the first business in Minneapolis to be electrified. When the blue limestone Pillsbury A opened in 1882, Minneapolis flour catapulted onto the European markets with brands still recognizable today, Washburn-Crosby's Gold Medal Flour and Pillsbury's Best XXXX.[4]

The rise in flour production was made possible by the rapid increase in the number of settlers plowing Minnesota's prairies and savannas for wheat cultivation. In 1880, ten million bushels of wheat were brought to market in Minneapolis; by 1882, that figure had increased to nineteen million.[5]

It wasn't only prairies and savannas that succumbed. The great deforestation of Minnesota's pine woods accelerated. In 1880, sawmills along the Mississippi River had produced 195 million board feet of lumber; in 1882, that number would expand to 312 million board feet, more than the circumference of the earth.[6] The lumber was essential to the growth of Minneapolis and St. Paul, and even more so to the settlement of the treeless prairie regions of the state.

WINCHELL REMAINED IN MINNEAPOLIS during the first half of the 1880s. The survey's initial foray into northeastern Minnesota had provided him with ample material to sort through, catalog, and identify. He supervised the production of hundreds of thin sections of rock from the North Shore and managed to scrutinize a third before writing the 1880 annual report.

By 1880, the survey had become much bigger than originally envisioned. The 1872 legislation had created a geological and natural history survey, and Winchell attempted to give equal weight to plants and animals, not often with success, since the legislature had also requested him to give first priority to geology.

In his trips to southeastern Minnesota, Winchell had amassed a huge number of fossils, particularly brachiopods, small invertebrates with fan-shaped shells, collected from limestone beds. He worked through identification of dozens of species, including several that were new to science. During this spell in the lab, Winchell was immersed in taxonomic details of shells—their dimensions, hinges, and interior scars left by muscle attachments. It was the kind of precise, detail-oriented work at which he excelled. By 1882, he was preparing maps and manuscripts on the geology of the state's southern counties.

Lastly, he was at work on a report of rock suitable for construction, as requested by the legislature, and he had begun assembling a final report on the survey's work in southern Minnesota, which would become volume 1 of *The Geology of Minnesota.*

The 1880 field season had Winchell colleague Warren Upham in southwestern Minnesota, investigating counties buried by a hundred feet of glacial till. While there, Upham also defined the borders of the Coteau des Prairies,[7] a prominent glacial moraine rising from its grassy surrounds. A portion of this is today called Buffalo Ridge. Later, Upham traveled north, to work out the extent of Glacial Lake Agassiz, which once filled the Red River valley.[8]

Winchell hired Oscar Garrison, a civil engineer from St. Cloud, to follow the drainage of the upper Mississippi River through Wadena County, the White Earth Reservation, and Lake Itasca, then downriver to Aitkin. By canoe and over portages, Garrison passed Ojibwe villages collecting Seneca snakeroot to use medicinally and to trade; the plant contains an expectorant used to treat respiratory ailments. Later, he came upon abandoned gardens four or five acres large near Lake Itasca,[9] and still later, many miles of forest blackened by fire, which he believed had been intentionally set by Ojibwe to draw in moose and deer.

The Reverend C. M. Terry, who had been on several survey collecting trips, compiled a comprehensive report on the state's hydrology for the 1880 report. He showed an unusual appreciation of aquatic ecosystems. "Leaning from the side of a boat on a calm day in summer," he wrote, "one may feast his eyes on little and delicate forms of beauty growing in miniature forests and jungles, where the larger bass and walleyed pike love to lie in cool and shady seclusion." He also valued the "thousands of beautiful lakes," which provided a respite from "mental and physical strain."[10]

In addition, Dr. P. L. Hatch, a Minneapolis homeopath, drew up a list of Minnesota birds, and university student T. S. Roberts supplied a synopsis on the state's winter birds.

The survey's annual report for 1880 ran over four hundred pages—this, after the board of regents and the legislature had ordered the annual reports to be "brief and synoptical."[11] Winchell held to that admonition for exactly two years (1878's annual report was 123 pages; 1879's was 183 pages) before reverting to his methodical, comprehensive self.

When Winchell told the regents that he had devoted "his whole time" in the 1881 season to the final report, the annual report, and the building stone report, that did not mean he planted his posterior firmly to his desk chair. In June, he revisited the limestone quarries at St. Peter, Kasota, and Mankato and the granite quarries at St. Cloud. He paced the blocks of downtown St. Paul and Minneapolis, carefully observing which stone buildings had sandstone footings and which had limestone, and which used Kasota stone for walls and which used brick. He noted that a few buildings employed "artificial stone"—concrete.[12] In his pocket field book, Winchell drew a grid. Of Minneapolis's estimated over eight thousand buildings, only about 3 percent were of stone; St. Paul claimed a slightly higher percentage.

WINCHELL MAINTAINED AN OFFICE on the third floor of the main university building, although by the 1880s he no longer taught classes. He had taught six years in the 1870s, and he did not miss it. Without classroom duties, he devoted his time to the survey. In his early forties, his dark hair receding and his bushy beard still unruly, he remained a vigorous presence on campus. *Ariel,* the student newspaper, commented from time to time on his activities and whereabouts: He entertained the graduating senior class at his home in 1882; he traveled to Washington, D.C., on "geological governmental business"; and when volume 1 of the final report of the survey, *The Geology of Minnesota,* came out in 1884, *Ariel* claimed it was "a monument to the industry, skill and ability of Professor Winchell and his assistants."[13]

Winchell's regular presence in Minneapolis benefited his home life, since the large Winchell brood was as active and complex as it would get. In March 1881, Lottie gave birth to a fifth child, Louise, in the family home at 120 State Street. That same year, the eldest, Hortie, spent the summer in the Red River valley as a field assistant to Warren Upham, and then entered the university as a freshman at age fifteen. He chose a scientific course of study; his entry exams had been unremarkable, except for math scores, echoing his father, who several decades before had been a standout in math at the University of Michigan. The dark-eyed, dark-haired Hortie, who resembled his mother, was not an academic grind. He joined a fraternity, played baseball (notable for his frequency of strikeouts[14]), and refereed Field Day events. He had inherited the Winchell voice, sang in the university glee club, and played a leading role in musicals.[15]

The next year, Ima, age fifteen, followed her brother to the university. Her entry scores were higher than Hortie's, perhaps revealing a more conscientious attitude toward both classwork and testing. She continued to resemble her father, with a broad brow and big blue eyes, a notably pretty girl. She embarked on a liberal arts curriculum and joined Delta Gamma Fraternity,[16] where she met fellow freshman Gratia Countryman, who matched her in intellect and in aspiration.[17] They soon became close friends.

Gratia, small and vivacious, caught Hortie's eye. Sometime in the early 1880s, the two became an item. Now a blossoming romance heightened emotions in the Winchell house on State Street.

Ima, petite and blond, set about mastering Latin and French (thanks to her childhood tutor, Benedict Juni, she was already adept in German), history, and rhetoric. She took an early interest in editing and, her mother's daughter, was something of a feminist before equality for women ever had a label.

By 1882, the Winchell house included these two college students along with Avis, age ten, and Alexander, age eight, students at Marcy public elementary school, and Louise, a one-year-old.

With five children, including four in school, the Winchells were familiar with childhood illnesses. Since their marriage in 1864, Newton and Lottie had kept a record of illness detailing various ailments and mishaps that befell family members. They recorded everything from births to remedies for rheumatism. Tellingly, Louise's arrival in 1881 got a one-sentence mention; Hortie's birth had elicited pages of excruciating detail. The Winchell record of illness showed two intelligent people trying to learn from experience in an age without antibiotics or modern theory of disease.

One by one, the Winchell children fell ill to scarlet fever, chicken pox, diphtheria, measles, and mumps, often infecting each other. One child, Alex, seemed remarkable in his maladies. He nearly died from dysentery the summer he was one. From then on, Alex suffered one illness after another. He had worms; a treatment designed to rid him of the parasites caused him to convulse. When Avis brought home measles from school, Alex developed a more severe case. When laboring under respiratory infections, he seemed to develop croup. In all of these illnesses, Newton assumed a nursing role as often as Lottie; he had been and continued to be a very attentive father.

But there were incidents neither parent could nurse. Tussling with Avis in the upper hall of the house, Alex stumbled and thrust his hand through a window, gashing the flesh and exposing tendons. A doctor was called to sew it up. When he was eight, Alex jumped from the family buggy hurrying to get out, slipped, and broke his arm. Another doctor was called to set the bone, which healed nicely. Horsing around with Hortie in the front yard, he fell (or was thrown by his older, stronger brother; the account was unclear) and broke the same arm. Again, a doctor set and cast the bone. Carelessly stacking firewood as Louise stood nearby, Alex threw a stick that grazed his little sister's head and split the skin. Newton was away from home for that accident; Lottie closed and dressed the wound, which left a scar.[18]

His jacket flapping, his stride purposeful, Newton Winchell heads toward campus. The August morning is unseasonably cool, a subtle foreshadowing of fall. Bur oaks cast lacy shadows on the grassy lawn. Off in the distance, a large white tent, a "pavilion," billows in gentle whiffs of wind. Near the pavilion, the bluish stone of the main university building, topped with its signature cupola, presides over the campus. It is a fine structure, worthy of any to be found back East. He wishes that the science hall, only in the planning stages, were complete and available to the visitors, who will flood the campus this day.

The 1883 annual meeting of the American Association for the Advancement of Science, AAAS, is under way and in an hour or so, hundreds of conference goers will mill about, sorting themselves into geologists, chemists, biologists, paleontologists, ready to attend plenary sessions in their respective fields.

Winchell has prepared extensively for this national conference. As secretary to the executive committee, he has been a kind of glue, binding city officials, campus professors, scientists, and other participants into a functional whole.

His protégé, Warren Upham, who is also his boarder at 120 State Street, will address the geologists today, unveiling the state survey's research on the Minnesota River valley. He will identify for the first time the glacial river that carved out its oversized valley as the River Warren, named to honor General G. K. Warren, who initially worked out its hydrology.[19] *This meeting's geology papers focus on glaciation and its southern extent in North America. The survey certainly will contribute new ideas on that.*

The Winchells are also hosting Charles Hitchcock of Dartmouth, Upham's advisor, during the conference and, further, have invited the entire geology contingent to their house that evening. When Newton left this morning, Lottie was fussing with last-minute details and instructing the hired girl how things were to be done when guests arrive.

Arriving at campus, Winchell pauses. His mind is full of a million reminders of tasks to see to. Tomorrow will be a breather: the conference goers have a day at Lake Minnetonka.

IN 1883, THE AMERICAN ASSOCIATION for the Advancement of Science held its annual meeting in Minneapolis after being invited by the Minnesota Academy of Science and other Minneapolis civic groups.[20] Founded in 1848, AAAS was and still is the premier science professional organization in North America. Alexander Winchell was instrumental in its organization, and Newton regularly attended meetings and served on committees, although neither brother held a high leadership position in 1883. AAAS drew members from the United States and Canada and aimed to be universally inclusive by holding its annual meetings throughout the continent.

The event was a huge undertaking. A thousand attendees and spouses were expected, straining the capacity of Minneapolis hotels. Conference participants had the option of staying at Lake Minnetonka and taking the train in. Trains were affordable, direct, and efficient and ran frequently. The newly opened Lyndale House, built on high ground on Lake Calhoun's eastern shore, also had rail service and touted flush toilets, gas lighting, and steam heat, the latest in luxury.[21] Well-heeled attendees had this option.

Civic leaders pondered how to feed such a crowd. Participants were issued red ribbon badges displaying their registration numbers. Badges admitted visitors to the pavilion for noon dinner each day.

Communications presented new concerns. The post office at Bridge Square downtown routed attendees' mail to the university. Telegraph and telephone facilities were headquartered in Old Main. Social and personal telegrams were free to conference goers, courtesy of Western Union.[22] The executive planning committee delegated communication issues to prominent Minneapolis citizen William S. King.

In fact, it appeared that most of Minneapolis had a hand in the planning. There were nine committees, involving more than a hundred people. The list of committee members reads as a Who's Who in Minneapolis. George A. Pillsbury, father to the miller and brother to John S., led the executive committee and held planning meetings in his office at Pillsbury headquarters. Dr. Folwell, president of the university, headed the invitation and reception committee. Mrs. John S. Pillsbury, who had spoken vigorously for women's suffrage when Lottie had been elected, had charge of the Ladies' Reception Committee. Lottie was on that committee.

Then there was the matter of entertainment. It was not exactly a red-light district crowd. The executive committee planned a Saturday "pic-nic" at Lake Minnetonka (no scientific meetings that day), free tours of the flour mills on Saturday morning, and free rides to Lake Harriet to experience the famous Minneapolis lakes.

The geology sessions on glaciation were of particular interest to Minnesotans. The last plenary session featured John S. Newberry, chair of the Geology and Paleontology Department at Columbia University and Newton's erstwhile boss on the Ohio Geological Survey. His talk, "The Eroding Power of Ice," listed four "heresies" being bandied about: that there had never been an ice age; if there had been a climate change, it was warm and not cold; that evidence of "glacial action" was actually effected by icebergs; and that ice has little eroding power. Newberry refuted all of these, declaring, "What we want is facts, and more of them."[23]

A newspaper account of the speech reported that Newberry had barely finished speaking when geologist J. P. Lesley, recently elected president of AAAS for the coming year, leaped to his feet in disagreement, declaring there was not a shred of evidence that glaciers had any role in forming lake basins.[24] Minnesota's geologists must have listened with interest. This was Upham's field of active research. (Glaciers, of course, had everything to do with forming lake basins.)

The conference goers and their guests embraced a closing excursion to the scenic St. Croix River valley gorge and its geological potholes. The committee had expected fifty people for this trip; 250 showed up at the depot. The planners successfully wired the railway in White Bear Lake for additional train cars but were unable to reach Taylors Falls, where fifty people were anticipated for noon dinner at the Dalles House.

The train arrived at the gorge at eleven o'clock, and somehow the masses were fed. Guides from Taylors Falls took tourists (many of whom were geologists) on pothole tours. When the sightseers boarded the train for home, most had "their pockets full of rocks."[25]

Before the conference's end, the Geology Section had voted Newton Winchell in as its president for the coming year. This national post positioned Winchell to shape the professional organization of geologists into a form to his liking in the next few years.

IN 1884, VOLUME 1 OF THE SURVEY'S final report, *The Geology of Minnesota,* was released to the public. Winchell had finished the manuscript and submitted it to President Folwell in March 1882. The oversized book with forty-three plates consisting of county maps, illustrations of Minnesota buildings made from local stone, drawings of microscopic rock structure, and fifty-two figures of (for example) rock layers was a complicated publishing project. It was seven hundred pages. It covered twenty-eight southern Minnesota counties, eight surveyed by Winchell, three by Mark Harrington, and seventeen by Warren Upham, which represented 22 percent of the state's land mass.

The first section covered a history of Minnesota beginning with the earliest European explorers and their maps. Winchell researched the topic himself. He considered Father Hennepin "unscrupulous," identified the Dakota village on Mille Lacs in 1679 as "Kathio," and reproduced Jonathan Carver's sketch of the Falls of St. Anthony with a cluster of teepees on the western bank of the river.[26]

Winchell's extensive grasp of Minnesota history found its way into other publications: it was cited in the "Handbook of Minneapolis," given to the AAAS meeting attendees as the chief source for the background on the city,[27] and it served as the basis of a history of First Methodist–Episcopal Church of Minneapolis, of which Winchell was the historian.[28]

With the publication of volume 1, Winchell wrapped up the initial phase of the survey. He was now free to focus on the intriguing rock formations of northeastern Minnesota.

13

FIELDWORK, POLITICS, AND FEMINISM

1886–87

SIX FIELD SEASONS LAPSED before Newton Winchell returned to northeastern Minnesota. This time, he took a cadre of geologists and assistants, including his brother Alexander. In 1872, Alexander had left the University of Michigan for Syracuse University but soon resigned to accept a professorship in geology at Vanderbilt University. He left Vanderbilt after his writings on human evolution conflicted with biblical teaching. By 1879, he had returned to the University of Michigan, where he still owned his octagon house, and resumed his professorship in geology and paleontology. In that capacity, he joined the Minnesota Geological Survey in 1886.

Since Newton Winchell's last visit, iron ore mining had commenced on the Vermilion Range. A new town, Tower, named for one of the investors in the Minnesota Iron Company, sprang up on the eastern shore of Lake Vermilion. Railroad track was laid between the nascent mine and Two Harbors, a port on Lake Superior, and on July 31, 1884, the first 220 tons of ore arrived at the ore dock.[1] When Winchell and his team resumed work in 1886, iron ore had also been discovered in Ely,[2] and plans were under way to extend the railroad from Tower to what would become the productive Chandler mine at Ely.

Alexander had seen plenty of iron mines in Michigan, but he had never seen wild country like Minnesota's Arrowhead. During his three months of fieldwork, he marveled that there were no public roads at all from Tower on Lake Vermilion eastward to Ogishke–Muncie

Lake (off what would become the Gunflint Trail); there were no more than half a dozen white settlers and no Native American villages.[3] Yet native peoples regularly traversed portages between lakes. These paths were often deeply worn but always narrow.[4]

The exclusive mode of transportation in this region was by birchbark canoes made by the Ojibwe. Canoes were typically sixteen to eighteen feet long,[5] and Alexander's party and the other two survey parties used them to convey their camping equipment, their tools, and the ever-expanding canvas sacks of specimens.

Alexander found Tower a bustling town of two thousand, but "everything destined for [it] must be got over these portages. . . . The very first portage . . . is one of the most execrable in the region."[6] (His report to the legislature somehow overlooked the railroad line running into town.) However much he deplored the portage, he did remark on the presence of "exquisite" water lilies bedecking the clear waters he passed by.

Newton divided his men into three parties, and each hired an Ojibwe guide, who carried the lightweight canoe and served as cook. Alexander, heading a party, frequently consulted with his guide and kept a running record of Ojibwe names for the lakes; Burntside, near Ely, was "Ga-na-ba-ne-ia-bi-gi-teia-ga-ma." Alexander was much taken with the beautiful lake and predicted that "some day, the pleasure-seeker will discover [its] charms."[7] The Minnesota survey teams took care to use Ojibwe names for the lakes in general, but the smaller lakes were often unnamed, and it was Alexander who bestowed the name "Newton" on the small lake leading into Basswood Lake from Fall Lake, for his brother.[8]

Modernity in the form of mines now intruded into the pristine boundary waters. Out surveying an arm of Basswood Lake, Alexander reported, "We hear with great distinctness, the blasting in the iron mines. . . . very like heavy thunder from below the horizon."[9] The Tower mine was thirty-eight miles away.

IN SUMMER 1887, NEWTON WINCHELL took a party to northern Wisconsin and the Upper Peninsula of Michigan to inspect the Penokee and Gogebic Ranges. He noted their continuity with the rocks in northeastern Minnesota. Newton met Alexander in Sault Ste. Marie, then visited iron ore mines in Negaunee, the center of the Marquette

Iron Range, and the Bruce copper mines in Ontario, which they had visited on the Michigan Geological Survey twenty-five years before, when Newton was a youth, knowing nothing about rocks. The Bruce mines were no longer producing and displayed what Uly Grant described as a deserted town, a ruined environment, and slag heaps of crushed quartz left behind as waste.[10]

Back in the northeastern Minnesota woods, Winchell's teams—usually comprising one professional geologist, two assistants, and an Ojibwe guide—continued to collect specimens to complete a geological map of the region. These parties involved a variety of Winchells and Winchell relatives. Alexander led parties in 1886 and 1887; Hortie, who was an undergraduate in his early twenties and embarking on his own professional career as a mining expert, assisted and later assumed leadership of a party; and two future sons-in-law, Frank Stacy and Ulysses Grant (Uly),[11] undergraduates at the University of Minnesota, served as assistants. This was the opening chapter in Grant's illustrious career, which eventually took him to a professorship at Northwestern University.

Newton split his time between the woods and Minneapolis. In 1887, he lugged a heavy camera over portages, taking photographs of rock formations, but often he was not present with a team.

"Where is your father, anyhow?" Uly Grant wrote to Avis Winchell when in town at Ely. "I expected to have a letter from him, but did not get any. . . . half the men I have asked say that he went into the woods, and the other half say he went back to Minneapolis. The only thing I feel certain about is that he was here, as I saw his name at the hotel on the register. The postmaster don't know whether he is to hold mail for one or two Winchells."[12]

Grant's daily, lengthy letters to the sixteen-year-old Avis painted a vivid picture of geology camp life in the 1880s. Winchell had given each party a list of specific places to travel to and sample. Using Ely as a base for resupply, Grant's party paddled to Ogishke–Muncie Lake to collect seventeen sets of twenty-five specimens each. The work shredded their hands: "the rocks have a very disagreeable habit of cutting,"[13] he told Avis. The constant use dulled their hammers, so they frequently returned to Ely to have them sharpened.

There were other tribulations. Grant wrote ruefully of sunburned hands, necks, and noses and of the constancy of blackflies and mosquitoes. They looked forward to August, when they knew the bugs would drop off. Grant's party employed the same guides two years in a row: Pashitoniqueb and Augigeesick, who went by "Charley" and "Nick." The guides presided over the kitchen, catching fish, shooting an occasional rabbit (or once, a woodchuck, which they told the geologists was a rabbit—until they saw that the geologists found it delicious); and they carried the canoe, while the others took the packs of gear and rocks. Usually, there was so much gear they took each portage twice.

The guides had other skills. Once Grant's party was late in leaving Ely. Hortie had left earlier in the day, and Grant hoped to catch up to him, but Hortie had had a four-hour start on them, and soon Grant's party was paddling in the dark. "We paddled away until half past nine. I don't know how the Indian found the way, but he did without once going wrong. It was quite dark and there was no moon and he had been over that route only once before." The portage trail was rough, especially traversed at night, and the party finally decided to camp, without tents—Hortie had the tents. But they were so exhausted, they fell asleep immediately, and when Grant awoke, the moon was up and shining. He termed it "a perfect night."[14]

Grant and Hortie had been friends at the university, although Hortie was a year ahead in school. When the two parties met, they shared camp—playing cards, swapping novels, taking canoes out after supper to catch a picturesque sunset.

They restocked the camp larder at Ely. Grant, city-bred, was shocked by the new mining town: "This is a great town," he wrote sarcastically to Avis, "about five hundred inhabitants and twenty business houses, a good share of which are saloons; no church and no school house that I know of. About every man seems to smoke, chew, and drink; and added to these accomplishments, they possess the power of mixing something worse than slang into their speech. The stores are all open on Sunday and they do more business then than on any other day. It's a bad place, like all new mining towns. . . . I wouldn't want to live here very long."[15]

Tower, which was older by four years, was not more refined; the mine, four miles out of town, had a resident population of six thousand; Tower itself had two thousand people and two churches, one of which drew only twenty people the Sunday Grant attended. "This twenty," he observed," included the minister, the janitor and five little girls."

"A large number of people spend the day in a boat riding and fishing," he added, and "every other building on the main street is a saloon."[16]

BACK HOME IN MINNEAPOLIS, Lottie was experiencing success of a much different sort. Fervent in her Methodist faith and a strict teetotaler, she had been active in the Minneapolis Women's Christian Temperance Union (WCTU) since its inception in 1877, attending state-level meetings, often addressing attendees from the podium, and serving as a delegate to the national WCTU convention in Detroit in 1884. In summer 1887, she was savoring the sweetness of political victory.

Although today the WCTU is most associated with a hatchet-wielding Carrie Nation, at its onset the organization advocated broadly for interrelated social reforms, such as labor laws, public health legislation, curtailing prostitution, and education. It spoke out against domestic abuse and child labor, and it supported women's suffrage. Women who could vote would have a means to change the dynamics of their lives. Alcohol often played a role in domestic abuse and a disrupted home life.[17]

Lottie, college graduate, former educator, and wife of a scientist, was sought out by the president of Minnesota's WCTU to pursue the teaching of prohibition in public schools and colleges. When first asked, Lottie hesitated. "My heart at once responded to the call," she told the state WCTU convention in 1887, "but my hands were very full, and how to find the time and place . . . was a problem very difficult to solve."[18] Solve it somehow she did, and accepted the challenge in 1881, soon after Louise's birth, and with the four other children ranging in age from fifteen to seven.

While no longer on the school board, Lottie's experience both in the classroom and on a governing board gave her a unique view of what was needed. States were promoting legislation requiring the teaching of how the human body was affected by drugs, alcohol, and tobacco. Women around the country, wearing little white-ribboned badges, had toiled to get such legislation passed, and it had not been easy. One state, Rhode Island, had been able to secure passage without much WCTU involvement, and Lottie and her committee used Rhode Island's tactics as a model.

Enlisting sympathetic legislators, they drew up a bill and got it introduced in the house, where it was referred to a committee and subsequently languished. The women did not go to the capitol to advocate for the bill. They watched and waited from afar. When the clock ran out on the session, Lottie observed, "We had learned . . . that Success does not happen; it must be won."[19]

Two years later, in 1887, the women again pursued legislation, this time with clout derived from the national WCTU convention, which had just been held in Minneapolis. Armed with petitions, a letter-writing campaign, and friends in high places, the small dynamo Lottie had more success. She and other "white ribbon women" were in St. Paul at the capitol when the bill was passed. She was present again the next day when "we had the unspeakable pleasure of seeing our bill become the law of the State."[20]

That summer, she and other officers in the WCTU assumed the responsibility of inspecting textbooks for classroom use. Many of those serving on the committee were practicing teachers.

Lottie's work with the WCTU was not finished, and she remained active, even attending the international WCTU convention in London in 1895.

Newton Winchell opens the front door and lets himself in to 120 State Street. The large, comfortable old house is unusually quiet. Only the ticking of a clock can be heard. The reddening rays of a setting sun warm the room. It has been another long day at the university, but now he's home for the evening. The aroma of a pot roast wafts from the kitchen.

Little Louise runs in to greet him. Still in pigtails, only seven, she is the late-in-life child to bring him and Lottie additional joy. "Papa!" she cries happily.

"Where's Ma?" he asks, shrugging off his overcoat and hanging it on the coatrack. The first requisite of a properly managed household was knowing that Lottie was at the helm, running the show, knowing where the children were, what they had been doing that day, and who would be home for supper.

"I'm in here," Lottie calls from the pantry.

Newton walks down the hall and into his study and stares, aghast. Changes! It's been cleaned! Someone has tidied up the stacks of papers, moved his books about, volumes he had precisely opened to the exact page needed. His eyes narrow.

"Lottie!" he bellows as he tears out of the study and back down the hall. "The fiend's been around again!" The "fiend" is the cleaning woman, who from time to time would enter the professor's sanctum sanctorum at Lottie's request, and attempt to dust and polish. Newton's scowl is ferocious as he confronts his placid wife. She is not intimidated, though, even as he towers over her. Newton likes to bluster and make a big fuss, but he can't hide the merry twinkle in his blue eyes. It tells her the bluster is just for show.

THE HOUSE AT 120 STATE STREET continued to be a beehive of activity. Ima, in school year 1887–88, was in her senior year at the university and had taken on the responsibility of managing editor of the national Delta Gamma quarterly magazine and maintained a high profile as an ardent proto-feminist. ("I didn't expect a lecture on woman's rights . . . and I don't want your sister, or you either, to think so," Uly Grant wrote Avis.[21] Grant was a university classmate of Ima.)

Upon graduation, Ima was still romantically unattached, a fact the student newspaper made much of: "There seemed to rise in Ima's sky / no matrimonial star" ran the poem of the class of 1888.[22] Meanwhile, Ima, in her graduation essay, titled "Woman," claimed the focus to be not on "What can a woman do?" but rather on "What is there she cannot do?"[23] Nevertheless, she would marry classmate Frank Stacy the very next year.

Avis completed high school in 1888 and entered the university, pledging her sister's fraternity, Delta Gamma. The next year, she became an assistant editor at the organization's *Anchora,* following in Ima's footsteps. Avis was a good student and inclined toward poetry, but her heart was not in her studies. Before she even matriculated at the university, she was wearing Uly Grant's frat pin, and the two became engaged some time in her freshman year. Grant, five years older, was at the time pursuing graduate work in conchology (studying shells) under Winchell.

Alexander, two years behind Avis, happily toiled at the high school. In a shop class, he crafted a wooden bookcase to hold the family's twenty-six-volume dictionary. Newton hung the case on the wall just behind his place at the head of the dining room table—in easy reach to settle a family dispute that might arise during dinner over the meaning or spelling of a word. A family working as editors (Newton, Lottie, Ima, Hortie, and Avis all edited copy at various times) might well argue about words. Occasionally, Newton would pronounce the dictionary "wrong," and then there was arguing and discussion, as the challengers consulted other sources. Newton was usually proven correct.[24]

Louise, in what she would later term her "care-free" childhood, did not attend school.[25] Newton and Lottie thought that their older children had been under too much pressure in public school and kept Louise out of the classroom until she was ten—although she undoubtedly was tutored at home by both parents.

Only Hortie was no longer at home full-time. Entering his twenties, Hortie had acquired a wide range of experience in geological fieldwork. As a child, he had accompanied his father to the limestone cliffs in southeastern Minnesota. He had worked under Upham in the Red River valley at age sixteen, learning about glacial activity, and headed up survey crews in northeastern Minnesota in the summers of 1886 and 1887. In 1887, he transferred to the University of Michigan to work under his uncle Alexander. He would later claim that despite exposure to other fields, he was always only attracted to mining,[26] and he would publish with his father a treatise on the Mesabi Iron Range in 1891, a year before mining began on that fabulously rich deposit.

Despite his having given her an engagement ring, Hortie's romance with Gratia Countryman had run its course by 1887, and while he was in Ann Arbor, he became engaged to his first cousin, the strikingly beautiful and musically gifted Ida Belle Winchell, whom he would marry in 1890.

In 1887, Newton was immersed in the production of volume 2 of the final report of the survey. But there was more, so much more, occupying his thoughts as dreams for his profession of geology blossomed.

14

SHAPING A SCIENCE

1888–91

Newton Winchell moves through the house making a bedtime check. The windows are open to catch the southeasterly breeze. Curtains lift and fall, lift and fall, as cooler evening air wafts through the interior. St. Anthony Falls burbles in the distance.

The summer is aging. Already it is mid-August. The birds on campus have fallen silent, their familial duties completed. Crickets have taken over the night, creaking a steady rhythm. Minneapolitans are on vacation. The Sunday Minneapolis Tribune *ran a column of names of people fleeing to Lake Minnetonka to escape the city heat.*

But not Winchell. A momentous time is at hand for Minnesota's state geologist. Tomorrow he will board the train bound for Cleveland and the 1888 AAAS meeting. He has planned to arrive early, like the other geologists in AAAS, for he himself has issued a summons to his colleagues: the time is ripe to organize.

Winchell pauses before the big grandfather clock in the back parlor. His nightly routine involves winding it up, a pleasing ritual. The old clock—his Grandfather McAllister bought it in 1816—makes a distinctive whirr as the weights are reset, and with that whirr, the past rushes in, flooding Winchell with memories. He can see his Grandmother McAllister's parlor at Spencer's Corner, see her winding the clock as he now does, recall the time he spent as a callow young schoolteacher when he lived with her.

Winchell is now forty-eight. He seldom has the leisure to think about the past. This decade has been filled with demands of work and family that anchor him firmly to the here and now. But every night, just before bed, he rewinds the clock, allowing himself the pleasure of measuring how far he's come.

AT THE AMERICAN ASSOCIATION for the Advancement of Science (AAAS) meeting in Cincinnati in 1881, there had been talk of forming a strictly geological group and publishing its own journal. Although geologists had been instrumental in forming AAAS and had occupied its president's chair numerous times since its founding in 1848, many felt that the multidisciplined organization no longer met their specific needs. For one thing, AAAS convened in summer, the field season for most working geologists. It was both a nuisance and a loss of precious days to attend a conference in August. There were some who felt that AAAS, with parties and receptions on evenings and weekends, had become too social. And, too, many recognized that the number of geologists had proliferated and that these professionals would benefit from a monthly platform to air current research.

Section E, the geologists, discussed these issues at the 1881 annual meeting and appointed a committee to envision an organization of American geologists, one comprising researchers from both the United States and Canada. They designated people to confer with John Wesley Powell, director of the U.S. Geological Survey (USGS), to parse out cooperation between the newly formed USGS and state surveys like Minnesota, and to engage his support for a geological magazine.[1]

Change moved glacially slow, though. Geologists again discussed both a professional group and a journal in 1883, when AAAS met in Minneapolis, Winchell's home turf. But at that meeting, unexpected opposition surfaced. Loyalty to the bedrock AAAS remained fixed. Men who had sought change, including both Alexander and Newton Winchell, became discouraged.[2]

The next year in Philadelphia, no one even broached the topic. Then, sometime in the mid-1880s, opposition melted away. Deciding they had waited long enough, several geologists banded together to produce seismic change: a monthly professional journal, *The American Geologist,* whose inaugural issue appeared January 1888. In a published account of the founding, Winchell modestly did not name himself as instrumental, but *The American Geologist* most definitely was his baby.[3] Seven working geologists, chiefly from the Midwest, served as editors.[4] The first issues were published by the University of Minnesota's press.[5]

Throughout the 1880s, Winchell had chaired and served on various AAAS organizational committees. As chair of the committee charged with forming a geological society, he and the secretary of that committee, Charles Hitchcock of Dartmouth, ran a small notice in the June issue of *The American Geologist* instructing geologists to assemble on August 14 at 3:00 p.m., the day before AAAS convened in Cleveland. The notice also observed that a poll taken seven years before had registered 124 out of 126 respondents interested in such a group.[6]

In a run-up to the 1888 Cleveland meeting, Alexander Winchell traveled to Washington, D.C., to meet with John Wesley Powell. The USGS had been in existence only ten years. Many of the state surveys, including Minnesota's, were older. Prior to the USGS, there had been several federal surveys. They had not been coordinated and even had competed with each other. The USGS had been created to bring all geographical and geological activity, particularly in the West, under one umbrella.

Newton Winchell had looked favorably upon the USGS in its early years. He had communicated his support to its first director, Clarence King, in 1879, appreciating its ability to transcend state boundaries, providing scientific fact on which to base federal policy decisions, and noting that a federal agency would be able to "harmonize a great many apparent discordances" in surveys that had been done piecemeal.[7]

However, he expressed concern that a federal survey should not take over questions of local concern. State surveys were enthusiastically supported by a scientifically literate local populace—at least that was true in Minnesota. If such information gathering passed from local hands to the federal government, Winchell foresaw a less engaged, less supportive public.[8]

JOHN WESLEY POWELL, the second director of the USGS, was a rock star in the geology world. He had won fame in the Civil War, leading troops at the bloody battle of Shiloh, in which he lost his right arm to a musket ball. After the war, in 1869, Powell cemented that stardom by leading a party in open skiffs down the Green and Colorado

Rivers through the Grand Canyon, a portion of the United States that was unexplored and unmapped. Although ostensibly a scientific collecting trip, the expedition became a rough ordeal, rife with danger, running white water with drops of twenty to thirty feet, and climbing rocky precipices of the canyon rim—all with a one-armed leader.[9]

Additional trips followed that expedition, and in 1872, the *San Francisco Chronicle* reprinted in nine installments the journals Powell kept. The public loved it, and Powell's trip down the legendary Colorado River took its place in history alongside that of Lewis and Clark's explorations. The man was a genuine courageous hero, and if he embellished his exploits and dramatized his personal valor a bit, it was all part of the effort to popularize America's growing knowledge of the vast unknown.[10]

But a geologist Powell was not. He lacked any earned college degree, although after the war, he had been appointed to a professorship at Illinois Wesleyan University, possibly because of his father's connections to the Methodist church.[11] His collecting efforts were haphazard, resulting in a mass of specimens that remained uncataloged and unidentified.[12] He was adept at stirring the imagination of a public eager for a glimpse of the wonderful, wild West, however; and he became skilled at getting funding for future expeditions.

There was perhaps more than a hint of envy behind Winchell's observation to his brother that "Powell is a smooth politician and ambitious operator, rather than a candid earnest working geologist," yet he acknowledged that "these qualities have made the [U.S. Geological] survey grow wonderfully well."[13] But the methodical, conscientious Winchell saw Powell as a likely imposter and definitely a usurper of geological turf.

Earlier in the decade, when Minnesota's Geological Survey was actively engaged in fieldwork, Powell and his men came to Minnesota and "went all over the state." He hired University of Minnesota geologist Christopher Hall, who had worked under Winchell and still owed him a report, and "set him at work on the very rocks that I was engaged on," Winchell wrote to Alexander, still nursing his grievance five years after the fact.[14]

Winchell had written a stiff letter of protest to Major Powell. In it, he pointed out that he had initially cooperated with the federal survey, had shared data and loaned material. He was dismayed and outraged

Newton Horace Winchell, circa 1885. Courtesy of the Minnesota Historical Society.

Charlotte "Lottie" Winchell, circa 1880. Courtesy of the University of Minnesota Archives.

Alexander Winchell with wife, Julia, and children Jennie Carissima *(left)* and Ida Belle *(right)*, photographed in Ann Arbor, circa 1863. Photograph by G. C. Gillett. Courtesy of Hennepin County Library.

A student in Professor Alexander Winchell's study in Ann Arbor with the professor's dog "Curley," circa late 1880s. Courtesy of the Bentley Historical Library, University of Michigan.

Nicollet Avenue between Third Street and Washington Avenue, Minneapolis, circa 1874. Courtesy of Hennepin County Library.

Newton Winchell around the time he arrived in Minneapolis, circa 1872. Courtesy of Hennepin County Library.

Old Main, University of Minnesota campus, circa 1880. Courtesy of the University of Minnesota Archives.

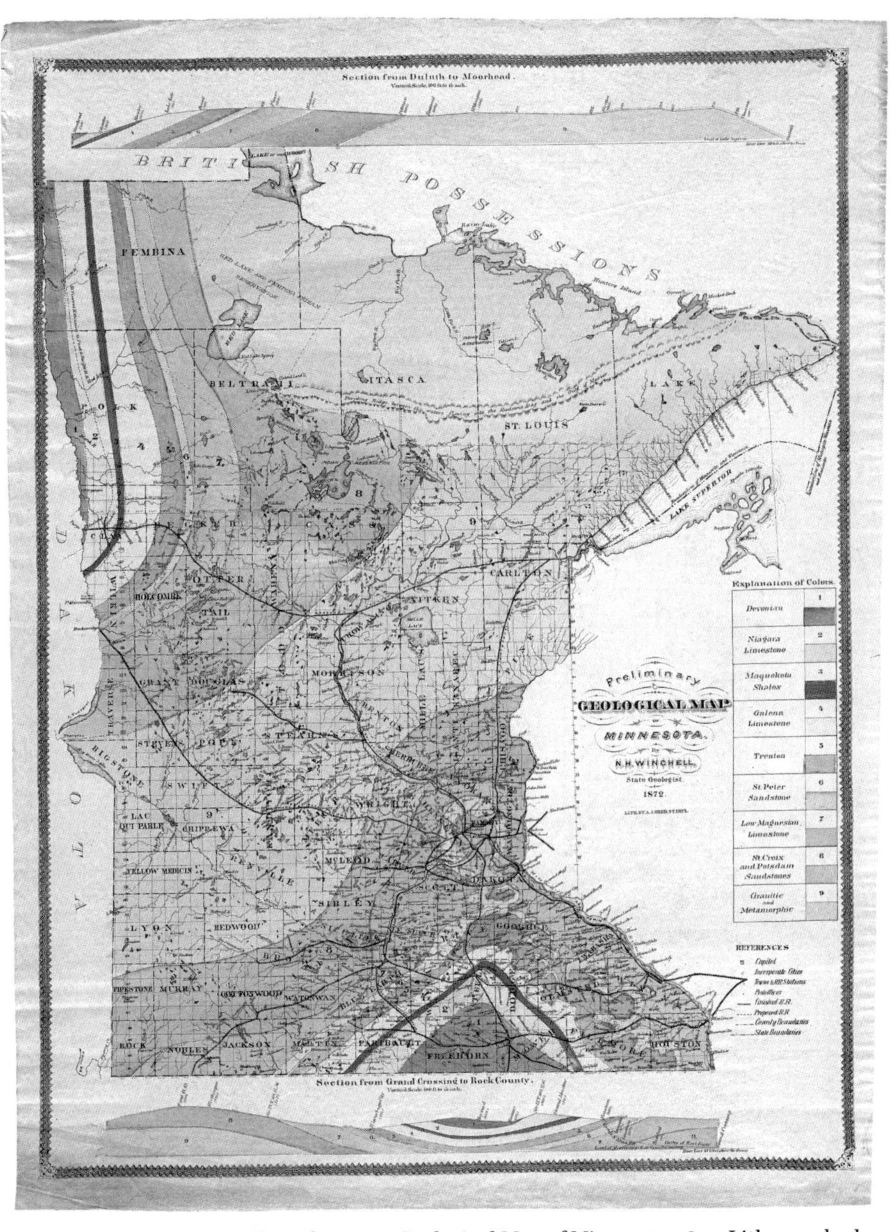

Newton Horace Winchell, *Preliminary Geological Map of Minnesota,* 1872. Lithographed by A. J. Reed, 22 × 17 inches. Courtesy of the John R. Borchert Map Library, University of Minnesota.

U.S. Army Expedition to the Black Hills, led by Lt. Col. George Armstrong Custer, photographed in 1874 by St. Paul photographer William Illingworth. Courtesy of the South Dakota State Historical Society, South Dakota Digital Archives (2015-11-30-314).

Spectre Canyon, Black Hills, photographed in 1874 by William Illingworth. Courtesy of the South Dakota State Historical Society, South Dakota Digital Archives (2015-11-25-303).

Winchell's field wagon on the western Minnesota plains. Photograph from volume 4 of Winchell's *Geological and Natural History Survey of Minnesota,* 1899. Courtesy of the University of Minnesota Archives.

North Sioux Falls Quarry, Eden, Minnesota, circa 1891. Courtesy of the Pipestone County Historical Society.

Castle Rock, a natural sandstone formation near Northfield, Minnesota, circa 1890s. Photograph by Ira E. Sumner. Courtesy of the Northfield Historical Society.

St. Anthony Falls, Minneapolis, circa 1860s. Photograph by Benjamin Franklin Upton. Courtesy of Hennepin County Library.

The Winchell home at 120 State Street, Minneapolis, the former Cheever Hotel. Courtesy of the Minnesota Historical Society.

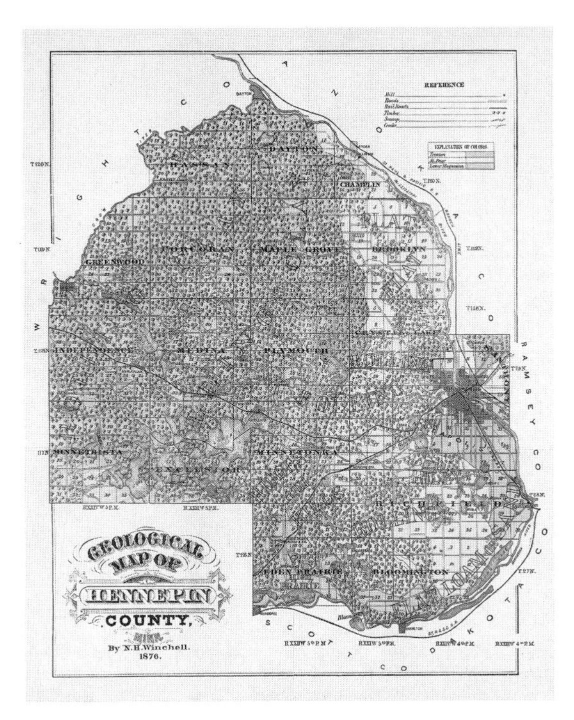

Newton Horace Winchell, *Geological Map of Hennepin County, Minn.*, 1876. 58 × 49 cm. Courtesy of the John R. Borchert Map Library, University of Minnesota.

Palisade Head on the North Shore of Lake Superior, circa 1875. Photograph by Bernard Freemont Childs. Courtesy of the Minnesota Historical Society.

Natural archway at the Point of the Great Palisades. Photograph from volume 4 of Winchell's *Geological and Natural History Survey of Minnesota*, 1899. Courtesy of the University of Minnesota Archives.

Winchell's birch-bark canoe used during surveys to the North Shore and Boundary Waters of northern Minnesota. The canoe is still in the collection of the Minnesota Historical Society. Courtesy of the Minnesota Historical Society.

Burntside Lake near Ely, Minnesota, circa 1885. Photograph by George A. Newton. Courtesy of the Kathryn A. Martin Library, Archives and Special Collections, University of Minnesota Duluth.

Fig. 1. Temporary Laboratory, Camp Overdone, Hoodoo Point.

Temporary laboratory at Camp Overdone, Hoodoo Point, near Tower, Minnesota. Photograph from volume 4 of Winchell's *Geological and Natural History Survey of Minnesota,* 1899. Courtesy of the University of Minnesota Archives.

Soudan Mine, circa 1885. This first iron mine in Minnesota opened in 1884. Photograph by George A. Newton. Courtesy of the Kathryn A. Martin Library, Archives and Special Collections, University of Minnesota Duluth.

Glacier geologist Warren Upham, circa 1880. He drew geological maps for western and northwestern Minnesota counties. Photograph by Jacoby & Son. Courtesy of the University of Minnesota Archives.

University of Minnesota campus buildings, circa 1900. *From left:* Pillsbury, Nicholson, Eddy, and Burton Halls; Old Main at far right. Courtesy of the University of Minnesota Archives.

Newton Winchell in his office at Pillsbury Hall. Photograph from 1893 issue of *The Gopher.* Courtesy of the University of Minnesota Archives.

The General Museum—Geology and Mineralogy.

Interior view of the General Museum of Geology and Mineralogy. Winchell articulated the large *Megatherium* skeleton. Photograph from 1897 issue of *The Gopher.* Courtesy of the University of Minnesota Archives.

The Winchell family outside the house at 120 State Street in 1894. *Back row, from left to right:* Alexander, Mabel Johnson, Francis Stacy, Ima Stacy, Horace, and Avis Grant. *Second row:* Newton, Antoinette Johnson, Charlotte (holding Alice Stacy), Julia, and Ulysses S. Grant. *Front row:* Ida Belle (holding Royal) and Louise. Courtesy of Hennepin County Library.

Exhibits inside the Galerie de Minéralogie, Paris, circa 1920s. Courtesy of the Bibliothèque National de France.

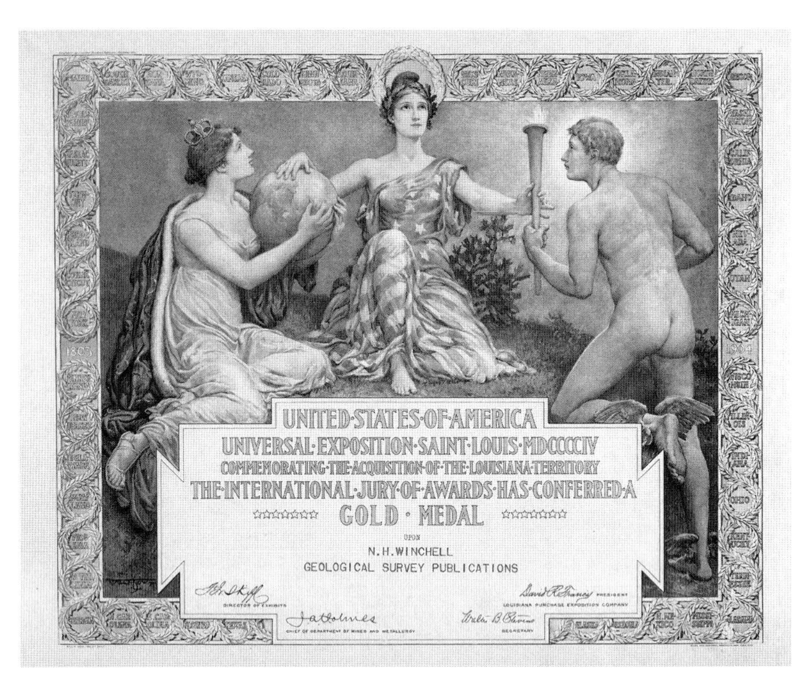

Gold medal certificate from the Universal St. Louis Exposition of 1904, presented to Winchell in honor of his survey publication work. Courtesy of the Minnesota Historical Society.

Newton H. Winchell Memorial Marker on West River Road and Franklin Avenue, Minneapolis, circa 1920s. Photograph by Arthur T. Adams. Courtesy of Hennepin History Museum.

that the USGS geologist, who had spent only months in the field, had arrived at conclusions that Winchell and Canadian geologists, working years on the oldest, or Archean, rocks, considered erroneous. Beyond feeling betrayed, Winchell was concerned that these wrong-headed conclusions would now bear the legitimacy of the USGS.[15] And he feared that his state survey would lose its funding.

The affront still simmered in his mind. "Powell treated me shabbily," he concluded to Alexander. "I gave him my objections [to USGS intrusion into Minnesota] and he never answered my letter—and I did not know until the receipt of yours that he even received it."[16]

Then, having gotten his resentment off his chest, Newton admitted, "These, however, are only some personal matters, and perhaps for the general good ought not to make any figure in the future."[17]

Reflecting on Alexander's reason to visit Powell in Washington, D.C., in the first place—to secure more amicable relationships between the USGS and state surveys—and knowing his brother's more polished temperament, Newton wryly conceded, "You were probably the proper editor to have a conference with Major Powell and the survey geologists."[18]

AMERICAN GEOLOGISTS MET in Cleveland's high school in August 1888 the day before the AAAS meetings opened. The gathering drew most of the country's prominent geologists, and after what *The American Geologist* described as "full and earnest discussion," they appointed a committee to draw up a constitution and bylaws for a geological society. It was tasked with securing membership and calling the first meeting. Alexander Winchell served as the chair. The committee decided on two meetings a year, one running concurrently with AAAS. The men retained dedicated loyalty to AAAS and didn't want the new society to be a break with it.

Those attending the meeting were termed the original founders.[19] Notably absent was Major Powell. He was AAAS's president in 1888 and undoubtedly had pressing last-minute obligations with meetings convening the next day. Powell was later selected for the first council of the new society, serving 1889–90.[20]

Before the meeting adjourned, there were thirty-three members. The Geological Society of America (GSA) became official at its next meeting, in Ithaca, New York, on December 27, 1888.[21] Henry Shaler

Williams, a paleontologist at Cornell University, had invited the geologists to Ithaca, and after the meeting, he and his wife hosted a party in their home.[22] By the deadline to receive "Original Fellow" designation, there were 112 names. From the onset, membership hailed from all over the United States and Canada and included geologists in federal and state surveys, and universities and colleges.[23] James Hall, one of American geology's grand old men, was elected the GSA's first president, although perhaps Alexander Winchell, as the most influential and active force behind a new society, ought to have been accorded the honor.[24]

Newton Winchell, who wrote an account of the meeting for *The American Geologist* in the February 1889 issue, ended with Latin *Pax nobiscum!*, Peace be with us!, harkening back to his youth in the classroom, teaching Latin, and signifying his sense that something momentous had happened.[25]

Two years later when the GSA convened at the American Museum of Natural History in New York City, membership had increased to 175 fellows. "It is evident," *The American Geologist* commented, "that the success of the Society is greater than could have been anticipated."[26]

Today, in 2019, the GSA boasts 26,000 members in 115 countries. It continues as the primary professional society for working geologists in North America. Its scholarly journal, the *Bulletin,* has been published continuously since 1890.

FOLLOWING QUICKLY ON THE HEELS of the establishment of both *The American Geologist* and the GSA, another major change in Winchell's life occurred: the opening of Pillsbury Hall in 1889, a large new building at the university devoted solely to science.

The new science hall represented the profuse flowering of Minnesota that took place in the 1880s. As Minneapolis's milling industry expanded and consolidated, workers poured into the state, and the city saw a building boom like no other. Flour brought great wealth to Minneapolis entrepreneurs, and in turn, they erected mansions, office buildings, and business blocks. In an effort to preserve some open space, the Minneapolis Park Board came into being.

Minnesota built a new capitol in 1883, its second. A four-story redbrick building with a central, open-air tower two hundred feet high, it was capped by a Romanesque dome. Housing all three branches of government, as does the present capitol, the senate chambers were lavish, with stained-glass windows, bird's-eye maple woodwork, and a magnificent gas chandelier. Alas, the state was growing so quickly that this elegant building would prove too small within the decade.[27] More durable in St. Paul were the massive mansions adorning Summit Avenue, although many of these, too, would be torn down for even bigger domiciles.

Across the river, Minneapolis would not be outdone by its twin. The Pillsbury A Mill, constructed of limestone quarried nearby, was completed by 1881. As businesses expanded, Nicollet Avenue extended farther south, pushing residential areas outward. Oak Lake, site of the present-day Minneapolis farmers market, was a fashionable subdivision with two small ponds and winding streets centered at Lyndale and Sixth Avenue North. The neighborhood at Tenth Street and Harmon Place, closer to downtown, also sprouted large Victorian houses. Minneapolis's lasting legacy of the 1880s boom would be the mansions lining Park Avenue extending south from downtown. William Washburn's Fair Oaks at Twenty-Second Street and Stevens Avenue went up in this decade.[28]

Of course, a mere favored few lived in mansions. Only blocks away from the aging but comfortable Winchell home, immigrant workers lived in a cluster of frame houses known as "Bohemian Flats," on the floodplain beneath the Washington Avenue bridge on the west side of the river. First settled in the 1860s, the neighborhood grew as Minneapolis boomed. The Winchell house, perched on the opposite bluff of the river, had a good view of the modest dwellings huddled along muddy dirt roads.

At the University of Minnesota, the boom took the form of rapid expansion. A new president, Cyrus Northrop, had assumed the reins of the school in 1884, and in rapid succession Eddy Hall (1886), Pattee Hall (1889), Pillsbury Hall (1889), and Nicholson Hall (1890) arose and shaped the campus. Heretofore, the university had consisted of two buildings, "Old" Main, which still dominated, and an agricultural building, which housed science and engineering labs and a greenhouse.

Eddy Hall, of red brick with sandstone trim, became home to mathematics and several engineering departments. Pattee Hall harbored the newly formed law school. Pillsbury Hall, constructed of two types of Minnesota sandstone quarried near Banning Junction and Fond du Lac, was designed to house the mineralogy, geology, and animal biology departments as well as a new school of mines. Iron ore was a hot commodity, and everyone saw a need to train mining experts capable of grappling with that industry in northeastern Minnesota. Nicholson Hall held the chemistry labs until a chemistry building was built, decades later.

The architect for Pillsbury and its siblings Eddy and Nicholson Halls was Leroy Buffington, working out of his Minneapolis office at Third Street and Nicollet Avenue. Buffington was a major architectural force in Minneapolis in the 1880s. He had designed the Pillsbury A Mill, the second capitol in St. Paul (a rare example of a Minneapolis architect receiving a commission in St. Paul),[29] the Tribune Building at the confluence of Hennepin and Nicollet Avenues in Minneapolis, and the ritzy West Hotel, also in Minneapolis. A draftsman in his firm, Harvey Ellis, was credited with creating the substantial Richardsonian Romanesque design of Pillsbury Hall with its rough-hewn stone blocks and hulking presence.[30] During the construction, a hollow where the students had maintained a skating rink in winter was filled in and graded for use as an athletic and drill grounds for the campus military program. Gopher football was coming into prominence. The skating rink had been a social gathering spot in past years, but there seemed to be little regret at its loss.[31] In 1940, the Bell Museum of Natural History would open on the former skating rink/parade grounds.

One wing of Pillsbury Hall was devoted to that forerunner of the Bell, the State Museum of Natural History, which had long outgrown its rooms in Old Main. The wing was to the east of the tower and termed "commodious" by Winchell—four times as large as the space devoted to the museum in Old Main.

Before Pillsbury Hall was even completely finished, Winchell and his staff moved the museum and survey material into the new digs, so that the space vacated in Old Main could be renovated for fall term 1889.[32] They labored diligently through the heat of the summer and the fall months, only to be set back by a fire occurring in the engine room in December. The fire did not destroy much material, but there was smoke damage and time-consuming cleanup. Winchell was frustrated by the lack of forward progress but assured his bosses, the board of regents and the legislature, that when the new home of the survey was fully equipped, and the new library stocked with reference books, the "nicer researches" still needing to be done on the extensive survey, in its nineteenth year, would go forward "with ease."[33] None doubted this. Winchell had proven himself indefatigable many times over.

As in its home in Old Main, the state museum held both natural history and geological displays. The university now employed a full-time zoologist, one of its own graduates, Henry Nachtrieb, and the responsibility of arranging the natural history displays fell on his shoulders.

Winchell (or, more likely, his assistant) set up the geology displays. Although these were chiefly Minnesota and other rocks and minerals, used by university students in their studies, one class of rocks captured the public's fancy and made headlines in local newspapers: meteorites.

Soon after Winchell ceased teaching in 1879, a spectacular meteorite fell in Estherville, Iowa, near the Minnesota border. Winchell dispatched astronomy professor E. J. Thompson to the scene to gather eyewitness accounts and a piece of the extraterrestrial rock. Thompson interviewed many people who described the event in apocalyptic terms. Glass windows in nearby buildings shattered, farmers working in their fields with horses were "stunned with fright," and one witness declared, "My soul! I thought the end of the world had come!"[34] The meteorite was huge. One piece weighed over four hundred pounds. Thompson managed to obtain a sample, which Winchell termed the most important acquisition of the museum in 1879.[35]

Then, shortly after the new museum in Pillsbury Hall opened, in a weird replay of the Estherville meteorite, another meteorite fell in northern Iowa. This time, Winchell sent Hortie, now serving as state assistant geologist. Hortie hopped the train to Forest City, Iowa, south of Albert Lea, Minnesota. Unfortunately for Hortie, the farmer who had located the meteorite wanted $100 for it. It was a big rock, over sixty-five pounds. Hortie counteroffered with $50 and was about to close the deal when a lawyer appeared and commenced a bidding war. Hortie dickered and finally prevailed. He hauled the meteorite to the train station in a wheelbarrow, but hours later, before the train left for home, the local sheriff confiscated the big meteorite and also numerous small pieces that Hortie had also pocketed.

The University of Minnesota state museum eventually got that meteorite, and Winchell listed fifty-eight meteorites in the collection in 1890, including some from as far away as Russia and Greenland.[36]

WINCHELL REPORTED in *The American Geologist* that the science hall had cost an initial $100,000 to erect but still needed an additional $100,000 to complete and furnish it. Why had the Minnesota legislature been so generous? Winchell attributed it to "enlightened public spirit which characterizes generally the communities of the western states."[37]

This spirit was forcefully demonstrated by steadfast university patron and former governor John S. Pillsbury, who, after the December 1889 fire, bestowed an additional $150,000 on the building to fireproof it and complete it.[38] The hall was named for its benefactor.

IN DECEMBER 1890, Winchell's mother, Caroline, wrote him and her other three sons, Alexander in Ann Arbor, Rob in Chicago, and Charlie in Washington, Indiana, beseeching them to alter Christmas plans and spend the holiday with her. She was eighty-four years old and living with her daughter Antoinette in Lynn, Massachusetts. Caroline felt it might be the last time she could gather her remaining five children together.

Alexander, the oldest, the son she had always turned to for money, for advocacy, and to help his younger siblings when she could not, wrote her affectionately and told her that he would come, if his brothers would also be there.[39] But he added that he did not think it would be their last meeting. Perhaps it seemed to him that Caroline was not that frail. Perhaps the indomitable woman seemed immortal.

Alexander was correct in believing his mother had a few more years in her. But he erred in thinking it would not be a last meeting, for it would be Alexander himself who would meet an untimely death in two months' time. Active and productive to the end of his life, Alexander died suddenly in February 1891 of aortic stenosis. With his death, Newton lost a brother, mentor, coworker, and friend. The loss was immeasurable. Newton now conducted the editorial duties of *The American Geologist* without a colleague to serve as a sounding board. Alexander had been at work on an extensive Winchell family genealogy; Newton assumed that work also.

A month later, Newton took out of his desk the account of the family grandfather clock that he had coaxed his mother into retelling. He dipped his pen in the inkwell and wrote, "The old clock stands in my back parlor, but is not kept running—indeed the pendulum is broken at the point of suspension and is removed. Mother still lives. Alex has left us recently."

15

THE AMERICAN GEOLOGIST

1888–94

In the early morning hours, before the sun has peeked over the treetops to the east, Lottie bustles about in the kitchen, feeding kindling into the stove, striking a match, and setting the coffeepot on to boil. Hearing her, Newton Winchell stirs on his makeshift bed in the study, groans slightly at his middle-aged stiffness, and glances at the clock on the bookshelf. Once again he had opted for the cot near the desk, rather than climb the stairs and sleep in a proper bed.

Twenty-four hours was not time enough for all the responsibilities he had piled on his plate. The American Geologist *needed constant attention. A deadline loomed each month to get an issue to the printers. He and Lottie were continuously reading text, correcting proof, and selecting illustrations for articles.*

He had successfully released two volumes of his final report to the legislature on Minnesota's geology. The third volume, part 1, wholly paleontological, was prepared for the printer. He thought maybe the days of brachiopods, bryozoans, sponges, and graptolites, all fossils of Minnesota's ancient limestone beds, were behind him. He and his team—Hortie, Warren Upham, and son-in-law Uly Grant—were pulling together their work on the remaining northern counties of the state.

The surfaces of his desk and drafting table reflected his workload. Here a stack of American Naturalists, *there a pile of reports from the Missouri Geological Survey. Papers in German. Papers in French. Letters from Washington, D.C., from Baltimore, from Philadelphia.*

"Newton?" Lottie calls from the kitchen.

Time to rise and shine.

CONCEIVED OF IN THE WINCHELL STUDY at 120 State Street, *The American Geologist* may have had many fathers but only one "lead parent": Newton Winchell. In 1892, the journal observed the fifth anniversary of its launch. "Are we sowing or reaping fruit?" the December issue editorial asked. "Too early to tell."[1] The editorial then proudly proclaimed that the journal was going out regularly to all parts of the world. Each monthly issue worked to bind the geological community together.

At the onset, seven working geologists, trained in the field and not in the lab, had agreed to function as editors. The seven originals were Samuel Calvin of the University of Iowa; Edward Claypole of Buchtel College, a small school begun by the Universalist Church in Akron, Ohio; Persifor Frazer of the Franklin Institute, Philadelphia; Lewis Hicks of the University of Nebraska; Edward Ulrich of the Illinois Geological Survey; Alexander Winchell of the University of Michigan; and Newton Winchell of the Minnesota Geological Survey.

While the midsection of America dominated the editorial team, the scope was wide-ranging. Issue after issue, the journal tackled geological inquiries on both coasts, north into Canada, and across the Atlantic, from Sweden to Italy.

THE EDITORS SET FORTH in their first issue the reason and purpose of the journal. They noted that there was no distinctly geological journal in the United States to publish and coordinate geological research. They wished to recognize the various geological labs and their publications—and to encourage conversation among scientists.

The format was set in the first issue and did not vary. Inside a heavy, brown paper cover and title page, scientific articles formed the bulk of the journal. These ranged from highly technical taxonomic discussions to descriptions of rock strata studied by a state survey. Not infrequently, the lead article was a biography of a notable geologist who had recently died. Newton Winchell had inspired this feature and wrote most of them,[2] although like the editorials, they were generally not signed. The biographies served to remind scientists immersed in arcane detail that the arc of geology was big and still in its ascendancy.

The articles were followed by editorial comment, sometimes short, sometimes lengthy, running on for many pages. A review of recent literature critiqued a vast range of publications. There were many state geological surveys in the Midwest nearing completion, and the editors summed up the volumes as they came off the presses. Many of the editors were bilingual or even trilingual (Winchell read both German and French), and their reports kept American geologists abreast of research taking place abroad.

A "recent publications" section simply listed new works released for consumption, and lastly, a chatty "Personal and Scientific News"—another unifier in the geological community—ran the gamut from promotions and job changes to facts about diamonds in meteorites or the discovery of gold somewhere. Early in the life of the *Geologist,* this "news" section reported that a "swindling geologist" was posing as a Russian savant (apparently convincingly)—or, strangely, a deaf-mute—and stealing fossils, microscopes, and books from trusting colleges and universities.[3]

In the first year, as point and counterpoint articles were published, the editors added a "Correspondence" section, running commentary letters from readers, often in full.

The founding editors realized that science teachers in classrooms, grade school through college, would benefit from a journal that provided them with solid geological information and how to present it, so there was a focus on education. And lastly, they pointedly observed that a general lack of cooperation existed among geologists. Many were suspicious of the USGS, especially those geologists heading up state surveys (like Newton Winchell). Consequently, they proclaimed *The American Geologist* to be a nonpartisan publication, running articles from USGS scientists and surveys alike.[4]

One goal the editors had not been able to attain in its first five years was to rotate the editorship, each editor being responsible for one issue in turn. Relying on the U.S. Postal Service, even on twice-a-day delivery such as Minneapolis had, they found it just too difficult to have articles submitted first to one city, then to a different city, so Newton Winchell had become the "first among equals," overseeing the compilation of each issue, copyediting, taking it to the printers, and after, readying each piece for mailing. After five years, *The American Geologist* was not yet financially self-sustaining, but it was close to breaking even.[5]

When money was short for the *Geologist,* it was even shorter for Winchell. He set the pattern in its earliest days, lending money to pay its bills—then being short for household expenses. In 1887, Winchell had extended himself to buy the family farm on Winchell Mountain in New York State, thinking it would be revenue neutral, with his Uncle Hilan paying rent. It had been a purely sentimental move, since Winchell had no possibility of farming it himself. But Hilan Winchell had not met a single payment, further strapping Newton.

In the first year, the *Geologist* had 218 subscribers, signing on for a year at $3.50 each. This had proved inadequate to meet costs. Early on, Newton applied to his older brother, much as he had as a college student in Ann Arbor. There is something eerily reminiscent of the teenaged Newton in the letters that began "Dear Bro." He had sworn then, as a teen strapped for money, that once he had an education and a career, he would never be in that position again. And yet, the "Dear Bro" letter continued, "The June *Geologist* will be out soon, but I shall have to [borrow] money to pay the bills, as I am shorter than short at present, and the *Geologist* funds are ditto."[6]

A month later, "I must have help. I can't pay my own bills, though I have my salary mortgaged (by loans) for several months ahead." Newton stressed that a lack of cash was not his fault: "It all comes from Hilan's failure to meet his engagements in the purchase of the farm on Winchell Mountain." And there were amplifying circumstances: "High taxes and unusual expenses all around have heightened the difficulty with my ready money," but Alexander would get his money back because "in a few months I may be easy again."[7]

Back and forth the letters flew between Minnesota and Michigan, each brother saluting the other, "Dear Bro," or if there was time to be leisurely, "Dear Brother." Often letters were sent daily, or even twice a day, as a deadline drew near. Newton informed Alexander what articles were running, if he needed more content in a particular section (maybe Alexander had something to add), and sometimes about his opinion on the validity of a certain article: "I am skeptical about Lake Erie having covered the site of Ann Arbor"—he and Alexander had both pondered the terrain around Ann Arbor many times as they rode the train out of town—"but his descriptions of those features are important"[8]

ALEXANDER HAD INFLUENCE not only because he was the elder but also because in his sixties, with grizzled beard and graying hair, he had accumulated luster as an educator. His passion was for teaching, and the earliest issues of the *Geologist* reflected this. Although his first paper for the journal dealt with Animike rocks in northeastern Minnesota,[9] the January 1888 issue also carried an editorial by him framed in Darwinian terms. "Geology in the Educational Struggle for Existence" laid out the case that geology, as a pure science (as opposed to medicine or pharmacy), remains perpetually underfunded because it does not convey a quick path to financial reward for its students. He lamented that school administrations were not "sufficiently informed" in the sciences to think of them as an equal means of culture as the study of Greek or Latin. Geology textbooks were at a competitive disadvantage because they needed to be illustrated and were more costly than English or Greek textbooks—and that was also true for zoology and botany texts. He hoped his editorial would awaken geology teachers to the disadvantage under which they labored, because often teachers were too focused on lesson plans and lab exercises to see the big picture. He hoped, too, to jog administrations into greater appreciation for a foundational science that had much to say about the nature of the world.[10]

The "Dear Brother" letters came to a swift end with Alexander's untimely death in 1891. So, too, did the conscious thrust toward advocacy of the geological classroom. Newton then scrambled to find his center of balance, even as the urgency of issue after monthly issue continued on.

THE AMERICAN GEOLOGIST WAS A convenient vehicle in which to publish papers coming out of the Minnesota survey. The Mesabi Iron Range was under scrutiny, and both Newton and Hortie—publishing under Horace Vaughn Winchell, F.G.S.A., a Fellow in the Geological Society of America—wrote papers and commentary. In a particularly prescient piece, "A Bit of Iron Range History," Hortie observed that "the average mining man has but little respect for the science of geology."[11] At age twenty-seven, as a young geologist, Hortie had heard whoppers told in mining camps of the incompetence of field geologists—no doubt, tales told specifically about his father, uncle, brother-in-law, and himself. He observed that discrediting geological science had several consequences: first, the miner wastes time, failing to discover his hidden wealth unless happening upon it by chance. Second, he wastes money by digging in places that science deems unlikely to yield results. And third, the geologist fails to receive credit for his work, because ore discoveries are considered accidents.

When the *Geologist* carried this opinion piece in 1894, the Vermilion Range in Minnesota was producing ore, but the Mesabi Range was still undeveloped. Hortie credited the first account of iron on the Mesabi Range to Minnesota's first state geologist, H. H. Eames, who explored the region in 1867; by 1894, he and his father had both established evidence of rich iron deposits there. But even today, the Minnesota Historical Society credits Duluth timber surveyor Lewis Merritt and his sons with "discovering" the Mesabi iron ore;[12] the scientists go unrecognized. What the Merritts found was deposits of iron ore of sufficient richness to be economically profitable.

Warren Upham, the glacier specialist who had done most of the work on Minnesota's glaciated counties, published frequently in the *Geologist.* Since wrapping up his work in Minnesota, he had joined the Geological Survey of Canada and continued work on the Red River valley and Lake Agassiz in Ontario and Manitoba. During his stint in Minnesota he had boarded with the Winchells. By 1888, he had married and returned to New England, his childhood home. He now lived in Somerville, Massachusetts, working on quaternary geology (glaciers) in that region under the auspices of the USGS. Upham's articles varied, sometimes reporting results of fieldwork, other times authoring comprehensive overviews of then-current understanding of glacial geology.

In 1894, the *Geologist* ran Upham's "Causes and Conditions of Glaciation."[13] In it, he reviewed current theories of what brought about ice ages, for in that decade, there was a growing consensus that there were a number of stages to the Great Ice Age, interspersed with warm intervals in which plants grew, plants like mosses and conifers.[14]

Today, geologists believe glacial periods are triggered by a constellation of circumstances. The position of continents with respect to each other and to the poles, variation in the earth's orbit, the amount of carbon dioxide in the atmosphere (which is influenced by plant cover), solar output, and ocean circulation patterns all play a role.[15] This complicated explanation is a result of decades of research and a refinement of various studies.

Americans, like Upham, noticing changes in elevation of glacial-till bearing regions, thought that significant uplift had brought on a cool, snowy climate. The Norwegian explorer Fridtjof Nansen suggested that a very slight change in the earth's poles could produce a glacial period.[16] And a civil engineer in California proposed that the Ice Age began because the earth was cloaked in water vapor that shut out the solar rays that warmed the planet.

The civil engineer defended his proposal against Upham's objections in a letter printed in a subsequent issue—nicely illustrating one of the strengths of *The American Geologist.* The journal had in a very short time established a forum in which scientists could debate and ponder dispassionate arguments, allowing time to digest and incorporate new ideas into existing bodies of fact and theory.

The many editors of the journal participated in this discussion with an extended "Editorial Comment" section. These were never signed, making it difficult to trace ideas to the scientist proposing them, but the anonymity was intentional, to allow a truly candid give-and-take. A geologist could take issue with an editor and duke it out in a letter published under "Correspondence."

And the geologist was always a "he." Even today, the field of geology is dominated by men. In the nineteenth century, a woman geologist was a rarity indeed. *The American Geologist* had, though, in its very first issue, reviewed a publication by a Dr. Mary E. Holmes, a new Ph.D. out of the University of Michigan, the first woman to be granted a doctorate in geology from that school, calling it a "lucid" account of fossil corals.[17] Later, it reported on a woman receiving a doctorate in inorganic geology from Johns Hopkins in 1893, noting that she was "the object of much curiosity on the part of the [local residents], who dubbed her 'the stone woman.'"[18]

Most of the published articles emanated from pens of geologists from the United States and Canada, but Europeans were regular contributors. Swedish geologists contributed their own research to the conversation on glaciers, although some of the research was not so weighty: the March 1894 issue announced that a Swedish paleontologist had discovered a bed bug in the graptolite shales in that country,[19] although from what we know about the evolution of insects today, that can hardly be true. The *Geologist* also published work done by Americans in northern Europe who were studying at European universities, especially in France and Germany.

In its first issue, the *Geologist* announced that it would welcome contributions from researchers with humble claims to fame. It kept this feature through its many years of operation,[20] and perhaps no author exhibited a humble origin better than Frank Bursley Taylor, writing at the time as a thirty-three-year-old independent researcher from Fort Wayne, Indiana, unaffiliated with any institution. Taylor's paper was a survey of abandoned shorelines of the south shore of Lake Superior.[21]

Taylor was notable for several things. First, he was a Harvard College dropout, leaving school because of poor health, quite possibly tuberculosis, which would explain why he was accompanied by a physician, a Dr. Pearce, on his field studies. His wealthy lawyer father footed the bill for young Taylor's expeditions, and Frank did not get a job until he was forty, when he was hired by the USGS as a specialist in glacial geology of the Great Lakes.

In 1910, Taylor would propose to the Geological Society of America that the position of the continents on the planet were not fixed but moved on the surface, and when they collided, it caused the uplift of mountains. He was not the first geologist to propose this. A number of scientists had looked for explanations that could explain, among other things, why the west coast of Africa seemed to fit neatly into the eastern edge of South America.

Taylor's mechanism for this "continental drift" was, however, original: he claimed that during the Cretaceous Period, the moon had been captured by earth's gravity. The moon's tidal pull dragged continents to the equator, in a kind of continental creep, resulting in the uplift of the Himalayas and the Alps.[22] Taylor's mechanism was eventually discarded for lack of evidence, but his claim that the continents moved on the surface was novel and startling and predated Alfred Wegener's continental drift theory by two years.[23] Wegener now receives credit for the proposal, which was not fully accepted until the 1970s and which radically changed the field of geology. Frank Taylor's work illustrates that young scientists not indoctrinated into the prevailing paradigm of their science are often those who make the creative leap into new ways of seeing.[24]

In the 1800s, geologists could not determine the absolute age of rocks or fossils, because radioactive decay had not been discovered. In 1902, Ernest Rutherford and Frederick Soddy reported that uranium turned into lead at a reliable rate; that is, the decay of uranium to lead could be used as a kind of clock. Determining the amount of lead in a rock compared to the amount of uranium could render an estimate of the rock's age.

Within five years, geologists had used radioactive decay to propose the earth was over two billion years old. This claim was mind-blowing—for a century, geologists had thought the earth was not much more than a hundred million years old.

Continental drift theory, many decades later, would likewise change very fundamental assumptions of earth scientists. Newton Winchell and *The American Geologist* laid the foundation to ready the science for these ground-breaking ideas.

The calm exterior of the big, shabby house at 120 State Street belied the whirl of activity taking place inside. Outside, snowbanks had begun to melt, and shadows on the frosty lawns were clearer in bright sunlight. The oaks now etched distinct silhouettes on the ground.

But within, a visitor would have met with a furious bustle. It was the end of the month, and the March 1894 issue of The American Geologist *was due at the post office. The long table in the sitting room was laden with stacks of the brown-covered journal. Wrappers waited labeling by hand, and after labeling, other hands folded protective paper around the magazine and swabbed a line of paste to secure it for mailing.*

The crew assembled at the table had done the mailing many times before. The American Geologist, *produced monthly, had been in production six years. Although highly professional in content, aimed at a North American audience, and boasting an editorial staff of twelve, the journal was directed entirely from Minneapolis. Newton Winchell selected submissions, wrote editorial comments, and reviewed recently published literature. Other editors contributed, but in the end he was the managing editor, making the calls on what to run and what to omit. Lottie was an indefatigable copy editor and would retain that role throughout the entire life of the journal. Rarely were there grammatical, punctuation, or spelling errors. When the issues were back from the printer and it was time to mail, all hands were on deck.*

The names of the workers that night are lost to the passage of time. Lottie was there. Newton, directing operations, for sure. Avis and Uly, who were living at 120 State Street, the family home's past history as a hotel making it easy to house extended family. Maybe a boarder; probably Alex, the geology undergrad; and certainly Louise, age fourteen, all pitching in, as one boarder remarked, "with friendly understanding and good fellowship."[25]

16

TERMINAL MORAINE

1892–95

IN 1892, THE MINNESOTA GEOLOGICAL and Natural History Survey turned twenty years old. At its onset, Newton Winchell had been a bright, vigorous thirty-two-year-old with a bushy beard, a thick thatch of hair, and a young family in tow.

Two decades later, on the tail end of middle age, his beard still bristled, his unruly hair was receding, and the family had grown into "a galaxy of erudites of such varied and acute mental alertness as taxes a father's tolerance."[1]

When he took the job, Winchell had been told it might last twenty years. Now, the geological portion of the survey was drawing to a close. Winchell's glacial effort, the full extent of his thrust, had been reached. Volume 1 of the final report had been published in 1884, volume 2 in 1888. Volume 3 had proved costly to print. It covered the paleontology of the state's Paleozoic and Cretaceous rocks and included numerous plates of fossils, from prehistoric sequoia leaves and sponge spicules to the many brachiopods and bryozoans found in the limestone layers of southern Minnesota. In the end, Winchell split volume 3 into two parts. Part 1 was published in 1895, but part 2, almost entirely gastropods (snails), was not released until 1897.

With the geological portion of the survey winding down, the surveys of plants and animals came to the fore. Professors at the university directed these: Henry Nachtrieb (zoology) and Conway MacMillan (botany). Liking editorial control, Winchell had asked that the annual animal and plant reports be turned over to him. The university's board of regents heard his case for this, then decreed that Nachtrieb and MacMillan report directly to them.[2]

A year later, the board further restricted Winchell's scope by voting to request that the U.S. government conduct a geodetic and topographic survey of the state.[3] As he worked to refine the geological map of northern Minnesota, Winchell had asked his field workers to take measurements by which to construct a topographic map of the iron ranges. The board of regents, anxious to bring to an end a survey that had outlasted its original source of funding—the sale of salt spring lands—acknowledged that the federal government could provide a topo map of the state, using surveyors that the state would not have to pay for. Winchell was ordered to cease.[4]

As a scientist, Winchell envisioned many different avenues of research branching from his baseline survey. In his annual report for 1891, he enthused that the third decade of the survey would show as great an advance of scientific knowledge as the first two.[5] In that report, he also noted that while he had planned to wind up the survey of northern Minnesota that season, he could not. The iron ore–bearing rocks were far too significant economically to end the survey without closer examination. He sent Uly Grant back into the field,[6] and he employed Hortie, now an established mining consultant, to write a comprehensive economic report of the entire Mesabi Range. The Mesabi Range was poised for rapid development. In the 1891 annual report, Hortie had given special credit to the Merritt brothers of Duluth and Oneota for their persistent search for good iron ore that eventually paid off, when all their friends had advised them to quit.[7] In the next annual report, Newton described one of the Merritt test pits and noted that the soft ore overlay a rock to which Newton assigned the name "taconyte,"[8] but later in the report he stated that the taconyte overlay the ore, calling the taconyte "a peculiar rock." He thought the ore might form as a result in a change of the taconite.[9] Unlike the ores of the Vermilion Range, iron-bearing rocks on the Mesabi were soft and near the surface. They were accessible by open-pit mining.

It was clear that Winchell's overseers, the board of regents, were not interested in further scientific study. They wanted the survey wrapped up. Winchell was deeply immersed in the emerging field of petrography. The wealth of undescribed rocks in northeastern Minnesota begged to be analyzed. The Winchellian penchant for methodically and thoroughly completing a task now rose up and propelled Newton Winchell to an unexpected place.

Although the field of petrography was advancing most prominently in Germany, Winchell was drawn to the French petrographers Ferdinand Fouqué and his friend and colleague Auguste Michel-Lévy. Fouqué had studied rock layers and volcanic activity. Fouqué had also introduced modern petrographical methods to French science, and he and Michel-Lévy had coauthored an important book on volcanic rocks.

Winchell's focus on northeastern Minnesota's granites, basalts, and gabbros, igneous rocks formed from magma arising from deep in the earth, paired naturally with French petrography. He began to think about traveling to France and working with those on the cutting edge of petrography, particularly Fouqué's student Alfred Lacroix, professor of mineralogy at the Muséum National d'Histoire Naturelle, a division of the Sorbonne.

Winchell could read French—he had taught it as a young schoolteacher—but the specialized vocabulary of geology challenged his knowledge. However, in the early 1890s, a young French-speaking medical student, Henry Beaudoux, rented a room from the Winchells. Years before, a German-speaking renter, Benedict Juni, had been employed to teach the young Winchells German; now, Beaudoux was engaged to help Winchell translate scientific papers and to brush up on his conversation.

In the summer of 1894, the two worked together to translate a monograph on rock layers in northeastern Minnesota, "The Extension of the Taconic Westward," to be read at the Sixth Annual Meeting of the International Congress of Geologists in Zurich, Switzerland, in late August. On steamy July evenings, Winchell and Beaudoux labored in the study at 120 State Street, discussing French idioms, grammar, and word usage. Laughter roared through the house as Beaudoux tried to explain double meanings to phrases Winchell attempted to use, though Beaudoux noted that Winchell was a "good French scholar."[10]

Soon French phrases crept into Winchell's writings. Greenstones were termed the *bête noire* of the geologists of crystalline rocks; a certain paper was called one of the *chefs d'oeuvres* of the U.S. Geological Survey.[11]

Professors at the University of Minnesota in the 1890s frequently went abroad for summer break or a semester sabbatical, often at half pay.[12] Winchell was not paid by the university, however, and European study did not conventionally fall under the duties of a state geological survey director. He hatched a plan to study in Paris; he would pay living and travel expenses—including those of Lottie and Louise, who would be fourteen—and call it a "vacation." He had not taken a single vacation in all the years on the survey.

In fact, he seemed to revel in a workaholic mind-set. When the student newspaper, *Ariel,* asked returning professors in 1891 how they had spent their summer, the question received a wide range of replies. President Northrop had had "a delightful time visiting Scotland, England, France, Germany and Switzerland." Professor Sanford had "enjoyed three weeks in Idaho." Professor Jabez Brooks had "roughed it" at Lake Minnetonka, and Dean Pattee of the Law School had spent a month in Boston. But Professor Winchell? "Didn't have any in the first place. Most of my vacation I worked in the laboratory."[13]

Now, that would change. Somehow, Winchell scraped together the money for three trans-Atlantic fares and arranged for son-in-law Uly Grant to supervise the survey and assume editorial duties of *The American Geologist* in his absence. Grant had recently defended his doctoral thesis on northeastern Minnesota rocks and was back in Minneapolis, living at 120 State Street with Avis. They would also provide a mature presence for Alexander, now age twenty-one, who would be a senior at the university.

Lottie would put her work with the missionary outreach of the Methodist–Episcopal Church on hold during her year in Paris. She was also deeply involved in the work of the Minneapolis WCTU, in charge of scientific temperance instruction. By happy coincidence, the International Women's Christian Temperance Union Convention would be in London in August 1895. Lottie was elected to serve as a delegate. Lottie's command of French was good; she could read and translate competently and probably honed her conversational skills with Mr. Beaudoux, the renter.

A fourth member would join the party headed for Paris: Newton's younger sister, Antoinette Winchell Johnson. Antoinette, who had made her home in Lynn, Massachusetts, after marriage and taught music at her husband's academy, was a recent widow. She had traveled to Minneapolis sometime after nursing her husband through his final illness in 1894 and lived at the Winchell home on State Street.

Antoinette had been Newton's favorite sibling growing up. She resembled him in looks, and, like him, had nurtured a burning desire to escape their precarious home life when she was young. With her music and her teaching experience, she had much in common with both Newton and Lottie. Like Lottie, she had stood for an election and served on the school board in her younger days. Antoinette would be a third intellectually curious American in Paris in the great adventure of 1895–96.

The warm summer breeze stirs the air and lifts the leaves of the Virginia creeper clinging to the clapboard face of 120 State Street. The Newton Winchell clan is gathered, everyone present, including the newly arrived baby, Royal, Hortie and Ida Belle's first child, dressed in a white flowing gown. There is a table groaning with food—Lottie has been cooking for days. A croquet course has been laid out on the lawn. The family's setter is trotting to and fro, his dog tags jingling from his collar.

The family entertains company—Antoinette Johnson and her daughter, Mabel, who are visiting from Massachusetts. Mabel plans to return home soon, but Antoinette will remain. 120 State Street is roomy! Antoinette is a recent widow, but she rustles about not in widow's weeds but a striped cotton dress with jaunty leg-of-mutton sleeves, in contrast to Julia Winchell, Hortie's mother-in-law and Alexander's widow, who still wears somber black.

The summer sun is fierce. The men, true to dictates, wear vests and knotted ties, but Frank Stacy, the populist, has loosened his. Damp tendrils frame the women's faces, dressed as they are in long skirts and sleeves. Ima and Avis are both pregnant.

Newton relishes his lively offspring, as he wields his croquet mallet and smacks the ball into the neighbor's yard. A straw boater protects his increasingly high forehead.

A photograph will mark the occasion. Is it possible to catch a moment where no one has his or her eyes shut? The photographer will try. Everyone assembles.

In departing, Newton and Lottie left behind the "galaxy of erudites," their offspring, now young adults engaged in myriad activities. Henry Beaudoux, the boarder, living in close proximity to the tightly knit clan, commented on their "varied and acute mental alertness," adding, "There were no dull moments in the Winchell family when they were awake, and it behooved one not to go to sleep, in order to ward off the blitzkrieg bombardments of challenges that were likely to flash from several directions, lest one be slain on the altar of ignorance."[14]

Ima, who had been the first to marry, was the first to make the elder Winchells grandparents. Alice Emily Stacy was born in October 1891. Ima's husband, Frank Stacy, who had once spent a summer on the survey and had discarded a scientific life for his formidable pen, had left his position as editor for the Howard Lake newspaper and now wrote for *Minnesota Magazine.* He was also involved with the Second Ward Populace Club, dabbling in political beliefs greatly at odds with his in-laws'.[15] Despite having a young family, Ima pursued feminist advocacy in the Women's Improvement League.

Avis had married Uly Grant in 1891 only weeks before the birth of Ima's daughter, and then departed for Baltimore, where Uly was finishing his doctorate in geology. Avis had not completed her university degree, a circumstance that must have distressed especially Lottie. By 1893, however, the couple returned to Minneapolis, and Avis proved in a high-profile way to be her mother's daughter. A married woman, she returned to the University of Minnesota as a junior to finish her bachelor of arts degree and cap it by graduating Phi Beta Kappa.

In November 1893, Avis made the local newspapers and the Boston-based *Women's Journal* for her activist role in the advocacy of "Dress Reform" for women.[16] In the 1890s, women still wore waist-constricting corsets, skirts to the floor, and sleeves of massive proportions. But times were changing. Women of the decade participated in social sports, like golf and tennis, and especially in bicycling, which had swept America in a frenzy. Avis participated in a "Dress Reform" forum overseen by mathematics professor Maria Sanford. Addressing an evening gathering of undergraduate women, various student speakers raised salient points. One noted "the evils of tight dressing, the compressed organs, habits of wrong breathing, and poor circulation." A man pointed out, "We are getting too fast to be hindered by the cut of our clothes," and asked, "Why do women try to torture themselves into a shape which no woman had by nature?"

Avis called on the university women to organize to further the cause. She noted that college-educated women became natural leaders, in fashion as well as in other areas. They needed to join with others to change fashion norms.[17] Later, she led the university's movement and demonstrated the ideal, modeling a "skirt [that] fell straight within four inches of the floor, and was worn with a long jacket with a soft front of blue silk." A newspaper reporter reassured readers that "the garment was in appearance feminine at all points save that of length."[18]

Hortie, the eldest of the tribe, was living in southeast Minneapolis with Ida Belle and his widowed mother-in-law. Introduced early, from the ground level, so to speak, to the ores of the Mesabi Range by his work on his father's survey, Hortie had recently gone into a partnership as a private mining consultant, after working for an early mining company on the Mesabi Range.

Ida Belle had inherited the Winchell voice (as had Hortie). European-trained, she was in demand in Minneapolis as a soloist, performing numerous concerts.[19] Sometimes Hortie joined her, his baritone harmonizing with her mezzo-soprano. Avis, an accomplished pianist, often accompanied them.

Alexander Newton Winchell, the fourth child, had his nose to the grindstone at the university. He had cast his lot early with geology. As a graduating high school senior in 1892, his address to his classmates had been on "Iron as a Factor in Civilization." At the university, unlike Hortie, he seemed not to have participated in organized sports but did resurrect the Chess Club, which met weekly at the Winchell home.

Seven years younger than Alexander, a true "caboose," Louise would not stay behind in Minneapolis—but her chickens would. The young teen maintained a flock and claimed that she could communicate with them. Newton found this both amusing and intriguing and enjoyed the interchange, half believing that Louise had actually mastered "chicken-speak."[20]

Rocks, music, women's issues, populism, chess tournaments, chickens. Louise—who undoubtedly was often held to "observer status" as the youngest—would later reminisce that at mealtimes "we always had very lively discussions on various subjects. . . . Arguments carried us far afield . . . subjects allied to Geology were most apt to enter into the discussion, and all of us without knowing it absorbed quite a bit of general information of a scientific nature." She added that they learned the meaning of terms, proper spelling, and grammar—all invaluable, since everyone was involved at times in proofreading *The American Geologist.*[21]

The year 1894 produced a bumper crop of Winchell grandchildren. In June, Hortie and Ida Belle became parents to a baby boy, Royal; in October, Ima and Frank Stacy welcomed their second daughter, Charlotte; and on New Year's Eve, Avis and Uly became parents to their first child, Addison, born at 120 State Street.

During the winter of 1894–95, a deadly influenza virus stalked Minneapolis. The Stacy grandchildren fell ill and recovered. Seven-month-old Royal Winchell was not so fortunate. The blue-eyed blond baby died shortly after the New Year. Hortie and Ida Belle were disconsolate. They would not have another child.

In the midst of this winter of joy and tragedy Newton and Lottie prepared for a year in Paris. They packed trunks and satchels. Newton filled a crate with rocks from the Arrowhead. He also prepared for *The American Geologist* a series of papers on the survey's work in northeastern Minnesota. They arranged for passage on the *City of Paris,* a notably fast steamship that landed in Southampton, England. From there they would catch a ferry across the Channel.

On Easter Sunday, as warm breezes chased the last of the sad winter away and Minnesota's landscape greened, the party of four boarded a train for New York. A wider world was waiting.

17

AMERICANS IN PARIS

1895–96

WHEN HE WAS SIXTEEN, Newton Winchell observed with satisfaction that he had already seen the great metropolis of New York City, largest in the United States.[1] As a young man, he'd had aspirations, but he had never dreamed of this: being roused from slumber by the tintinnabulation of Notre Dame Cathedral's great bells wafting over the River Seine as Paris stirred itself early on a Sunday. In late middle age, however, Newton had imagined larger worlds, and now, here he was, in the heart of France, embarking on great adventure.

The journey had begun on April 17, 1895. The Winchell party of four—Newton, Lottie, Louise, and Antoinette Winchell Brown, who would not stay the year—stood on Pier 14 jutting into New York's Hudson River. Their steamship, *City of Paris,* departed at 9:00 a.m.

The American Line's *City of Paris* had a reputation for speed. Under favorable conditions, she crossed the Atlantic in under six days. Three classes of passengers, each segregated to its own promenade, made the trip. First class sailed in luxury, with a vaulted stained-glass ceiling in the salon, lit by hundreds of incandescent lights at night, a ladies' drawing room, and a smoking room with red leather chairs. Second-class accommodations, located aft, were plainer but still comfortable; second-class passengers had their own dining room and deck on which to stroll. Third class, or steerage, was confined to the bottommost portion of the ship and was often filled with immigrants from northern Europe.

The Winchells left no record of how they traveled. It is hard to imagine the frugal family in the opulence of first class, still harder to imagine them in steerage. Assume, then, they traveled second class, walking the deck and taking substantial meals in the dining room—breakfasts of oatmeal, eggs, buckwheat cakes, and oranges; lunches at 12:30 of soups, meats, fish, and pastry; and dinners at six o'clock with multiple courses of soup, roast meat, game and puddings, pie, perhaps ice cream, and fruit.[2]

City of Paris docked in England's southern port of Southhampton. From there passengers transferred across the English Channel by ferry, and then via express train to Paris.[3]

Paris was alive and vibrant in 1895 when Newton and his little band of women arrived. The streets teemed with people, many of them foreign-born. Paris, in fact, had a higher percentage of non-natives than other European capitals.[4] The city was also a magnet for artists. An experimental school, the Impressionists, was gathering steam, and Paris was already a tourist destination for the Western world. The City of Light had weathered the Revolution, Napoleon's imperial reign, and decades of periodic war. In midcentury, Emperor Louis Napoleon Bonaparte, Bonaparte's nephew and heir, oversaw a massive urban renewal of the city. What had been an overcrowded, medieval snarl of decaying buildings and streets so narrow that not even a horse and cart could navigate, became over the course of several decades an elegant, spacious city.

The great reconstruction of Paris demolished sections of the city that had been unchanged since the Middle Ages. These high-density neighborhoods had been a source of disease, like cholera, and of urban unrest. The new plan replaced them with broad avenues that allowed light and fresh air into neighborhoods and eased traffic congestion. Green space, in the form of parks and public art such as fountains, appeared. Surrounding suburbs were annexed, broadening the tax base. Sewers were constructed, eliminating one source of disease. The renewal took place on both banks of the Seine, and thus the Paris that the Winchells encountered was one of rebirth and vigor.

They could not have missed the newest addition to the city's skyline: the Eiffel Tower. Built for a world exposition in 1889, the tower was on an entirely different scale than the renewed city and dominated the Left Bank of the Seine. Eiffel's tower had not begun its life universally loved. When planners of the exposition had imagined a commanding structure presiding over it, they had envisioned stone, not iron. One prominent architect considered Eiffel's structure "vulgar." As the massive structure began to rise over the city, prominent citizens mounted a protest, calling the ironwork a "monstrosity" and a "gigantic black factory chimney."[5] When the exposition opened, though, fairgoers agreed that the Eiffel Tower was the undisputable "star of the show."[6] Only six years old, the tower was engraving its image on the imagination of residents.

The Winchells took accommodations in the Latin Quarter near the natural history museum and the Jardine des Plantes, a formal garden that provided a small oasis in a very urban existence. Their apartment at 26 Rue Censier was new and likely reserved for visiting scholars at the University of Paris. The building fronted a narrow, curving side street, one in a row of seven-story apartment buildings with wrought-iron balconies and mansard roofs. From there it was a ten-minute walk to the Musée National d'Histoire Naturelle, where Winchell's primary contact, Alfred Lacroix, was based.

At thirty-two, Alfred Lacroix held the post of professor of mineralogy, a position he would have for the next forty years.[7] Photographs show him bearded with owlish spectacles in midlife, but he was only Hortie's age in 1895. Born into a family of pharmacists, he had made a tough decision to follow his heart into mineralogy. Lacroix had recently worked on volcanic rocks and was part of a pioneering wave of petrographers, a science concerned with the composition and properties of rocks. He employed methods he and his mentors had perfected,[8] and Newton had been drawn to these trailblazers.

The Musée National d'Histoire Naturelle, the French equivalent of the Smithsonian National Museum of Natural History, had its origins in a garden of medicinal plants maintained by King Louis XIII. This garden survived the Revolution and expanded to become a public park that boasted of a cosmopolitan array of trees and shrubs. Trees that were large when Winchell strolled the gardens can still be seen today. The brilliant minds of French natural history had long been associated with the garden, such as Georges-Louis Leclerc, Comte de Buffon (pre-Revolution); Jean-Baptist de Lamarck (straddling the Revolution); and Georges Cuvier (post-Revolution). Winchell undoubtedly knew of these French scientists, had read them in French, and perhaps had even used their research in his paleontological work. Lamarck and Cuvier were both experts in malacology (shells); Buffon had published a history of the earth and had proposed "Buffon's Law," which stated that despite similar environments, different regions had different species of plants and animals. This insight foreshadowed the science of biogeography, not then a recognized discipline. All of these men predated Darwin and varied in their views of evolution.

Winchell also had contact with two other geologists, who were mentors of Lacroix. Ferdinand Fouqué, sixty-seven, was the grand old man of petrography in France. He had worked on the geological survey of France and was the chair of the Collège de France. Fouqué had subsequently pursued research on volcanoes, working on the Greek island of Santorini. He had been Lacroix's thesis advisor, and rumor had it that he had refused to grant Lacroix his doctorate unless he married Fouqué's daughter—which Lacroix did.

Fouqué had collaborated with fellow geologist August Michel-Lévy on a groundbreaking work, *Minéralogie micrographique: roches éruptives françaises* (Mineral micrography: French volcanic rocks). Volcanic rocks, like all rocks aside from volcanic glass, are composed of minerals. These crystallize from molten rock deep in the earth, which cools and solidifies. If the magma crystallizes under the earth's surface, the process is slow, the crystals grow large, and the result is a coarse-grained rock. Conversely, if it crystallizes at the surface, it cools rapidly, and crystals are small, resulting in a fine-grained rock. Different crystals compose different minerals that look, behave, and affect light in different ways. The two French petrographers pioneered the study of rocks in thin sections (optical mineralogy) using both polarized and plain light to identify and quantify the different minerals and thus describe the mineral compositions of various rock types.

Michel-Lévy, who in 1895 was considered the most influential French petrologist,[9] was working on thin sections of igneous rocks, in particular the granites. Based at the École Nationale Supérieure des Mines de Paris (Paris School of Mines), he was systematizing the various igneous rocks and took a special interest in the plagioclase feldspars, very common minerals of igneous rocks. Tall and brilliant, he was nicknamed "Moses" for both qualities and was already something of a legend at age fifty-one.

Into this small cluster of petrographical luminaries came Newton Winchell from the young and still sprouting University of Minnesota, hauling his crate of hewn rock samples from the Lake Superior Arrowhead wilderness.

WINCHELL HAD MASTERED THE ART of making thin rock sections and had taught himself some petrographical methods, but in France he could learn state-of-the-art techniques and interpretations. Winchell was interested in identifying the mineral composition of the Minnesota rocks; Michel-Lévy was in the process of drawing up methods of working with crystalline rocks that were different from procedures in favor in the United States, where German methods dominated.

One scientific article coming out of Winchell's Paris laboratory dealt not with northeastern Minnesota rocks but rather with a meteorite that had fallen in northwestern Minnesota in 1894. Dubbed the Fisher meteorite, the extraterrestrial rock had excited much interest at the time. The farmers who had found it had immediately broken up one of the two pieces that had fallen intact, hoping to find gold.[10] Winchell determined the meteorite was chiefly olivine, a heavy, greenish mineral, alas far more common than gold.

The University of Minnesota Board of Regents, overseers of the Minnesota Geological and Natural History Survey, weren't interested in the chemical structure of rocks. They had always been primarily interested in the economic uses of the state's geology. They deemed the survey work completed. Winchell, who all along had seen the survey as a scientific endeavor, was not done, though. He had not fully described the composition of Minnesota's igneous rocks nor had he explained the conditions under which they had formed. In coming to Paris, he had chosen to satisfy intellectual curiosity and feed his compulsion to be thorough. He framed the year as "a respite" from "the tedium of office duties and their interruption."[11] But time away would not be without its consequences.

Newton Winchell pushes open the heavy entry door to 26 Rue Censier and enters the bustle of a Parisian morning. The sounds of the city—iron wheels rolling over flagstone, human feet on pavement, dinging bells, the nasal tonality of French being spoken, the coo of urban pigeons, the chatter of chaffinches high in the trees—all have become familiar in the past six months. Housewives with baskets on their arms pass him, heading to the market street of Rue Mouffetard.

Winchell turns left and treads the hill to the University of Paris, toward its natural history complex of buildings, all in creamy limestone. Autumn has come to the streets of Paris. There is a decided nip in the air—mild, though, compared to Minneapolis—and the chestnuts are turning a lovely shade of brown. Today, the heavy coat that he packed feels good.

The neighborhood is full of young people. Students stride by on their way to classes. Women are admitted into the University of Paris, and, in fact, in this very neighborhood, a brilliant young woman had just that summer married her lab colleague, Pierre Curie, and taken his last name. Marie Sklodowski of Warsaw, Poland, had left her native land when denied entry into its universities. She had been nurtured at the University of Paris and would become its first woman professor. Later, she would be awarded not one but two Nobel Prizes—for the discovery of radiation (1903) and for the discovery of radium and polonium (1911).

Academics lost in thought or chatting animatedly to colleagues join the throng toward the university. It is not unlike the University of Minnesota. Winchell feels at home.

At the intersection with Rue Geoffroy-Sainte-Hilaire, Rue Censier turns into Rue Buffon. Winchell crosses Rue Geoffroy-Sainte-Hilaire, walks through the big iron gates into a courtyard paved in brick. Big, old chestnuts shade the space. He turns right. An ancient Japanese pagoda tree guards the entry into the solid Minéralogie Bibliothèque built of graying limestone. He pauses for a moment to admire its massive neoclassical columns with plain Doric capitals and a triangular pediment over the entry. It's the pediment in particular he finds interesting. Two fat cherubs wedge books between them. One holds a sphere, the Earth, extending it toward a man, who sits pondering. His chin rests in his left hand. His right hand probes the orb, and where his index finger makes contact, it is cracked open, exposing its secrets.

Newton Winchell is that man.

LOTTIE, TOO, HAD WORK TO DO in Europe. When she learned she would be in France in the summer of 1895, the Minnesota Woman's Christian Temperance Union elected her a delegate to the World WCTU Convention in London in June.

Shortly before the summer solstice, Lottie traveled to London, possibly accompanied by Antoinette. The meetings were in central

London in the fashionable Marylebone neighborhood at a new concert venue, Queen's Hall. The hall could hold three thousand people, and its acoustics were superb. It would quickly become the premier concert hall in London and was home to the famous promenade concerts ("the Proms") that ran from 1895 to 1941. But in summer 1895, it held thousands of women from around the world. Lottie was thrilled to be there. Fervently committed to ending the destructive power of alcohol and other addictive drugs, and a deeply spiritual Methodist, she wrote a detailed report of the proceedings of that week to the Minnesota WCTU.[12] Lottie conveyed the facts: how many women delegates were there (nearly four hundred), who all the speakers were ("the peerless leaders of the world's women"), and what songs were sung (among the many, this one: "There Are Bands of Ribbon White around the World").

The hall housed a stunning organ. The delegates sang rousing hymns. The main stage held pink geraniums and white spirea. Two balconies encircled the ground floor, and an ornate ceiling soared upward. The illustrious leaders of the WCTU in the United Kingdom and the United States addressed the women, who had traveled from as far away as Burma and Australia to attend, an indication of the power of the WCTU in the closing decade of the 1800s. A member of Parliament spoke to them about another worrisome addiction. The opium trade in China, he said, was a "burning disgrace to England."[13]

At the end of the convention, Lottie was swept away once again by a closing finale at Royal Albert Hall on the other side of Hyde Park from Queen's Hall. This was a truly mammoth hall, seating ten thousand. The delegates and sympathetic WCTU supporters filled it. An eight-hundred-voice "White Ribbon Choir" presented music to the many different civic groups in attendance, including Good Templars, nuns from the London Temperance Hospital, and girls from an orphanage. There was no building, no exhibition hall of this scale in Minneapolis. The color and the pageantry overwhelmed Lottie. She filled her report with superlatives.

After the convention, Lottie continued to send reports back home to the WCTU periodical *The White Ribboner,* published out of Duluth. In October 1895, she wrote at length about the relationship between the French and red wine. She had read about the French understanding of the beneficial effects of wine, nay, even of its necessity, and she claimed she had "tried to lay aside all previously formed opinions." (Her friends may have doubted that was possible.) After several months in Paris, Lottie stated emphatically that "everybody drinks intoxicants: men, women, children, everybody." One could not avoid it. "Bottles of liquor are upon the table, in hotels, in boarding houses, in restaurants."[14] The phenomenon was most obvious in cafés, which differed from American saloons, as far as Lottie could ascertain, only in that they were patronized by both men and women; were wide open, so everyone could see people drinking; and often spilled out on to the street, where men and women sat with glasses filled with ruby contents, on view by all. Those in the know told Lottie that the only dry place in Paris was the WCTU headquarters.[15]

But drinking was not confined to eating establishments. On a drive down the Champs-Élysées, "the most beautiful avenue in the world," "the lovely gardens on either side were thronged with people, men and women, elegantly dressed, seated under the trees, talking, drinking." Lottie was shocked. Shocked. She had been told to expect this, but she had no idea. Out sightseeing, she had asked for water, and the guide told her no—it would make her sick. Would she like a little wine? "I most persistently refused it," Lottie declared in her dispatch home.

In her report to the Minnesotans, she concluded that wine is a "monstrous vice fastened upon this people" and "the whole nation is led captive" without anyone appreciating the dangers. "Where is the Moses," she asked, "to lead this nation against their will, out of a worse than Egyptian bondage?"

SADLY, MORE DISPATCHES DETAILING LIFE IN PARIS did not leave Lottie's pen, for sometime soon after she fell and broke her arm, and letters to home ceased until February. By then, what had once been an "alien land" to her had become home. Louise, who had enrolled in a Parisian academy to improve her French,[16] was fluent in the language and thriving. Her parents must have tutored her somewhat, for she read Shakespeare's challenging *Richard III* that winter, surely not a pleasure read for a fourteen-year-old, and surely she did not read it in French.[17]

Time now began to fly. The Winchells feared that they would not get to see all that was on their list of sights. They desired to see the

Chamber of Deputies, the elected body of representatives. France was the first country to adopt universal suffrage—for men, of course, not for women.[18] Professor Lacroix managed to get entrance tickets for them.

The chamber met in a neoclassical building and was supposed to convene at two in the afternoon. Newton and Lottie arrived promptly ("Tell Mrs. Rhame I was in time once," Lottie wrote to Alex back home). But the hall was empty, and an usher told them that things would start happening around three. Newton, not wasting a minute's time, whipped out a French scientific paper and began to read it.

Lottie looked around. She noted that the front of the hall was of white marble. Newton glanced up. "Carrara marble," he told her.[19] She observed that there were no "filthy spittoons"—as there were in the Minnesota state capitol. The French may drink, and they may smoke, but they never chewed snuff. They also didn't smoke in the hall, lighting up only in one designated corridor.

Newton was not impressed with the assembly. He thought the men looked "inferior," balding and gray in their remaining locks.[20] (As opposed to him, whose forehead was merely rising, and whose locks held as much brown as gray, perhaps.) Lottie was not impressed either. The chamber finally convened an hour and a half late. It hardly could be called to order. Noisy and inattentive, the members seemed to disregard the little silver bell that stood for a gavel. The representatives groaned and hissed at speakers, exclaimed "Mon Dieu!" and pounded their desks. Lottie thought back to the orderly, respectful way the World WCTU had conducted business in London and wrote to Alex, "but then, we were only women and these were men, who understand how to rule, you know."[21]

As spring came to Paris, Newton took a day off from the lab, and he and Lottie spent the better part of an afternoon at Père Lachaise, the renowned Parisian cemetery, situated on a high plateau overlooking the city. What drew the two to this place—years before British playwright Oscar Wilde and American musician Jim Morrison were laid to rest there—they did not say. In 1896, Père Lachaise was the most famous cemetery in Paris, with brick-lined avenues running like diminutive streets past the lavish monuments. Conifers bordered the throughways. Lottie noted there were enough English and Americans buried among the French to warrant an "Avenue Anglais."[22]

The Winchells admired the various monuments to the dead: mausoleums, chapels, slabs, pyramids. Unlike most sightseers at Père Lachaise, they took special note of the stones put to use. Limestone was most common, but marble could be seen in columns and statues. They identified granite, both red and black,[23] and some green trap rock.

Père Lachaise was not anywhere near the oldest cemetery in Paris. The city had established it in 1804, less than one hundred years before the Winchells' visit, but there were already space limitations, and Lottie wrote Alex that graves were laid atop each other. In their strolling, they came across the new crematorium and columbarium (such a new idea to Lottie that she lacked the term to describe it). In the columbarium, the spaces to house ashes were eighteen inches square. She approved of the idea: "a great saving of space."[24]

It was not a morbid outing for the two. Plants were blooming, birds were singing, and nature seemed full of life, not death. After all, those two Methodists believed in a Resurrection and "Him who burst the bonds of death and triumphed over the grave."[25] They were hungry when they got back to the apartment, and Louise had dinner waiting.

THE WINCHELLS DEPARTED PARIS in April 1896. They had been abroad exactly a year. Newton had arrived with a bum ankle sprained in fieldwork six months before. He left with it nearly healed. Lottie's arm had mended, and she had seen enough fashion that before she left, she brought her silk dress to a dressmaker to be restyled, to wow the Minneapolitans back home.[26]

Their travel plans took them back across the Channel to England, Louise quite seasick and everyone happy to land. They immediately caught a train to London and took a room on Russell Square, close to the British Museum. Newton made a beeline to the museum the next day to see the geological collection and meet a friend.[27] Lottie and Louise walked to Madame Tussaud's Wax Museum in Marylebone. They spent the day there and enjoyed it thoroughly, much more so than rocks.

The next day was Sunday. London was the birthplace of the Methodist Church, founded by John Wesley, a circuit rider who traveled on horseback across England, preaching several times a day. His chapel on City Road, London, still in use today, drew in these faithful three.

A statue of Wesley stands beside the church, and the house where he died is next door. They also strolled through the adjacent cemetery, site of the graves of such Christian notables as John Bunyan, Daniel DeFoe, and Isaac Watts.

In London on Sunday, shops were closed. This was not true of Paris—"a wonderful contrast," Lottie wrote, approvingly.[28] This meant, though, that dinner was hard to come by. They settled for an afternoon meal of toast and tea, cold ham, eggs, and bread and butter. Refreshed, the trio then made their way to St. Paul's Cathedral, for their second church service, evensong, far more ritualistic and formal than the Wesleyan service. It was the Easter season, and they were treated to the cathedral's big bells and magnificent organ.

The year abroad was over. Winchell's examination of northeastern Minnesota's rocks had been thorough and satisfying. As all sabbaticals should, his had rejuvenated him. Buoyed by this and the exotic, foreign life that they had heretofore only read about, Newton Winchell, with Lottie and Louise, boarded the steamship and headed home.

18

GLACIAL RETREAT

1897–1905

WITHIN A MONTH OF HIS RETURN from Paris, Newton Winchell met with a member of the University of Minnesota Board of Regents, which oversaw the geological survey. The board had tolerated what must have seemed to them a long delay, but now they wanted a firm date for its completion.

Winchell told them that he thought he could finish volume 4 of the final report in one year's time and had then "bent every energy to that end."[1] But he did not make that deadline.

He had had trouble making deadlines. In pulling together volumes 1–3 of the final report, he had needed to return to certain locations to reexamine particular rock formations. He had even sent a new team to examine the Rainy River gold district straddling Minnesota and northwestern Ontario. Popular opinion was excited by the prospect of gold on the border. Winchell felt compelled to provide professional assessment. The examiners—Hortie and Uly—had deemed Minnesota's portion of the gold district meager. The mine at Rainy River had operated a mere fifty-two days.[2]

Now, in 1897, Winchell told the board that the fieldwork for geological mapping was finished. However, volume 4 was proving rather unwieldly. He now saw it could be a volume as large as the first three volumes combined, and that was only the first two parts of volume 4. His assessment didn't include the microscopic work he had done in Paris.

To celebrate the completion of fieldwork and the survey's twenty-fifth year, Lottie threw a party in September 1897. She sent out invitations to all who had been involved with the survey, most of whom had been employed when very young. Many were now respected, established scientists. Invitations went out to Clarence Herrick, who had been the first hired in 1873 and was now the president of the University of New Mexico; Charles Schuchert, curator of the U.S. National Museum (forerunner of the Smithsonian); Edward Ulrich, who was then newly arrived at the U.S. Geological Survey; and John M. Clarke, a professor of geology at Rensselaer Polytechnic Institute. Lottie also invited the local Minnesotans who had so freely offered their help to the geologists who came to their county looking for rock outcrops. One of these was William Hurlbut of Rochester, who had been fifty-two when Winchell had met him in Olmsted County and was now in his seventies.

The old house at 120 State Street rocks with merry guests. The warm glow of gaslight spills from the windows and illuminates the Virginia creeper vines clinging to the clapboard. It had been one of those golden days of autumn, with warm sunshine and a gentle southern breeze. One of those days when folks know that the warmth can't last, that winter is on its way.

The day has given way to a delicious evening. The last of the sunset is a pink rim to the west, and the grumble of St. Anthony Falls can be heard in the distance. Voices of male raconteurs waft from the house, providing a basso continuo, punctuated by laughter: "It had rained for eight straight days. Mud was everywhere . . ." "So he said, 'where's my hammer?' . . . it was three miles back at camp . . ." "We were blown off Superior and then the sky grew weirdly dark . . ."

Newton Winchell, threading his way through the crowded parlor, takes it all in. The faces are so familiar to him. Some he sees regularly at geological meetings. Others he hasn't gazed on for years, and it disconcerts him to see these men, who had been so young when they worked under him, aging.

Well, he wasn't getting any younger. He didn't feel like he'd slowed down, but gray flecked his beard, which may have hidden jowls. The buttons on his waistcoat were getting harder to fasten.

On the sideboard, the punch bowl has been drained. As Louise squeezes past him, he stops her. "Louise, go tell Ma the table needs replenishing." He heads for the porch and some cool September air.

AFTER THE LAST OUT-OF-TOWN GUEST had departed and Lottie had set the hired girl to tidying the house, it was back to work for the boss. In particular, volume 4 of the final report, covering the geology of northeastern Minnesota, remained incomplete. Given that highly lucrative mines were producing astounding quantities of iron ore in the two mining districts, the Vermilion and the Mesabi, Winchell was most concerned to provide as much information to the state as he could organize and analyze. The first twenty-nine chapters were richly illustrated with almost ninety maps, photographs, and sketches, which needed to be engraved and reproduced.[3] This would be costly. In the end, there would be six volumes: volume 5, which Uly would write—essentially his Ph.D. thesis—and volume 6, an atlas of maps. Ever methodical, Winchell also wanted to write a final discussion, pulling all the research threads together, including the most recent findings.

He laid all this out in a December 1897 report to the board of regents. He observed that there needed to be a paper on the Winnebago meteorite that had fallen in 1890 near Forest City, Iowa. Thanks to Winchell's efforts, the university had a nice collection of these extraterrestrial rocks. His last annual report, number 24, also needed to be printed, and his Paris work, too, perhaps in the form of a bulletin.

In his report, Winchell also raised the matter of the museum, part of the 1872 legislation enabling the survey. Simply put, the museum was a mess.[4] He pinpointed the beginning of its disorganization to 1879, when he ceased teaching classes and Professor Christopher Hall assumed the job. Hall had full access to the collection and had requested samples of rocks to illustrate his lectures. Winchell loaned him survey samples. Later, Hall organized his own teaching collections, with a different numbering system than that of the survey. Over the years and under the care—or neglect—of two masters, the museum disintegrated. Hall made purchases without consulting curator Winchell. When Winchell needed, say, certain trilobites or brachiopods to write up reports, Hall denied Winchell access to them. But in his report to the board of regents, Winchell didn't blame Hall—Hall had wanted to do his job well. Rather, Winchell saw the mess as an unintended consequence of having two people and two separate concerns—university classes and the survey—utilizing the same collection.[5]

As to putting the museum to rights, Winchell, having so much other work on his hands, told the board, "I have not had the heart to attempt to rearrange and keep the museum in good order."[6] Further, he had lost his prep room in which to prepare museum displays. During the year he was in Paris, that little workroom—specifically built for the museum and connected to it by a special stairway—had been handed off to the head of the School of Mines. Winchell wanted it back. He also needed new display cases and new material.

"Pillsbury Hall is overcrowded," Winchell declared.[7] The board of regents already knew that. The massive stone structure housed geology, zoology, and botany and now the School of Mines, which had been formed without building space. Winchell thought that the museum should be given its own home, and he had a plan.

He thought the university should make plans for a museum strictly concerned with geology. It should display minerals, fossils, rock samples, and archaeological material, something that other top-ranking universities had. He pointed out that the Minnesota Academy of Natural Sciences, still vibrant and mostly driven by university scientists, needed a permanent home (they had been meeting downtown for many years at the city library at Tenth Street and Hennepin Avenue). A new building could house the museum, provide a meeting room for the Academy, and act as a hub for scientists from Minnesota and elsewhere to meet and exchange ideas. The geological library that he had amassed could be housed there as well. Lastly, postgraduate work in geology and mineralogy could also be headquartered in the new building.

Winchell painted a vision of an influential geology program of national, even international consequence. He had taken the program halfway there by his splendid, comprehensive survey, his bridging work in Paris, and his leadership on *The American Geologist.* He then pointed out that with the survey drawing to a close, he was free to direct his energies to the proposed endeavor. "If I can be instrumental, in the next twenty-five years, in establishing a geological museum, such as I think the University needs . . . I shall be satisfied," he told the board.[8]

The board of regents heard Winchell out and decided to address his issues and proposals at their meeting in April, four months hence.

After Christmas, Winchell, needing follow-up examination of the Arrowhead's igneous rocks, went back to Paris for six months. He was not home when the board met in April and declared that the geological survey was "substantially completed." They asked that Winchell complete volume 4 but that it not be published at present due to lack of funds. All assistants to the survey—meaning Uly Grant—would be terminated December 1898, eight months away. The five thousand specimens for the museum were to be classified, labeled, and displayed, but the workroom would remain in the School of Mines, because there was no other space. Winchell would need to wrap up the survey work by December 31, 1898, and the museum reorganization completed by June 1, 1899.[9]

On that date, at age fifty-nine, Newton Winchell would be out of a job.

THE NEWS DID NOT REACH WINCHELL until he arrived home in Minneapolis in June 1898. But first, he was greeted by familial drama that had occupied the Winchells for most of May.

Hortie's wife, Ida Belle, recorded the drama in a letter. Alex, age twenty-four, had become entangled in a love triangle. He had been engaged for some time to a local girl, Clare. But when Flora, his first cousin once removed (and Ida Belle's niece), arrived at 120 State Street for an extended stay, tension ensued. "Aunt Lottie thinks Flora is in love with Alex & that she is getting a great deal of influence over him," Ida Belle reported to Hortie, adding, "Uly thinks so, too."[10]

Events moved quickly that spring. Two days later, Alex had decided to end the stressful drama and marry Clare. "He does not love her & that's the worst part of it, he as much admitted it to me last night, but has an idea he must marry her because he promised to," Ida Belle told Hortie.[11] Lottie must have rued the fact that neither Newton nor Hortie, both of whom had broken early engagements in their impetuous youth, were at home to help him sort this through. Lottie didn't sleep that night.

The next letter to Hortie described "the most melancholy wedding I ever saw."[12] Ida Belle provided the music. When Lottie requested a lively gospel hymn, Avis, Ima, Louise, Frank, and Ida Belle gave it a go—Louise and Ima sobbing instead. When the hymn ended, Ida Belle looked around the parlor to see everyone in tears. So she launched into the wedding favorite, Mendelssohn's "Wedding March," and played it "with great gusto to try and cheer them up."

Alex and Clare then left on a honeymoon, which would take them to Paris and the apartment at 26 Rue Censier, where they would live while Alex took his doctorate in mineral petrography under Alfred Lacroix.

Newton left no record of his reaction to this familial storm. Flora was a favorite of his—his brother Alexander's granddaughter. Alex and Clare would create an enduring thirty-four-year marriage with five children. Decades later, after Clare's death in 1932, Alexander would marry Florence "Flora" Sylvester.

Hortie, receiving these three letters from afar, thought this must be an April Fools' joke.[13]

A FEW DAYS LATER, Winchell received the word that the survey would discontinue and that both he and Uly needed to look for work. "What's going to become of this family," Ida Belle wrote in dismay. "It will evidently be scattered to the four winds."[14]

When the board of regents' decision was released to the wider world, reaction in the academic world was one of disbelief. The faculty of the university signed a request that Winchell be appointed as professor in the geology department.[15] John Stevenson, president of the Geological Society of America in 1898, was flabbergasted. "Well, I'm absolutely sorry that your work on the state survey is so nearly done. . . . the state of Minnesota owes you a great deal & it should not leave you to seek another position."[16] Clarence Herrick, who had long experience with Winchell as a former student, employee, protégé, and fellow geologist, wrote an impassioned letter to the board of regents testifying to Winchell's prominence on the American scientific scene ("one of the best American petrographers"), his capacity for work ("a painstaking and honest worker with no patience for gush or display and possessed of a dogged persistence"), and, most important, his kindly nature ("He almost never had a word of disparagement of his colleagues even while he was suffering at their hands the most base and unscrupulous ill-usage").[17]

The local newspapers had their say. The *Minneapolis Tribune* merely reported that the board of regents had decided that the state no longer required a survey.[18] But the *Minneapolis Journal* declared, "It is difficult to account for the vote of the regents of the university for its discontinuance on any other ground than enforced economy . . . [but] the geological survey is the only department of the work carried on from the sale of the Salt Spring lands that can easily be shown to have paid for itself," and went on to advocate for a continued presence in the mining boom of northeastern Minnesota.[19]

NOW BEGAN THE SCRAMBLE FOR JOBS, a position Winchell surely did not expect to be in as his sixtieth birthday approached.

Harvard University had an opening, the Sturgis–Hooper lectureship. The position, an endowed chair, was then and continues to be one of the most prestigious in the natural sciences at Harvard. Winchell's contacts were uncertain of the requirements for the appointment. The preceding year, Hans Reusch, head of the Geological Survey of Norway, had occupied the position. Apparently, the plan was to offer it to a European scholar for a year's duration.[20] That would seem to disqualify Winchell. But Nathaniel Shaler, engrained at Harvard as the head of the Lawrence Scientific School, did not mention that. He did say the appointment to the position was not made by the university but by the faculty of the museum, making a fine distinction of which Winchell was possibly unaware.[21]

Despite letters of recommendation from active, prominent geologists, it was hard for Winchell to crack the Ivy League glass ceiling. The position went to William Morris Davis, a Harvard alumnus, not a European, who held the chair for fourteen years.

After this, Winchell set in motion his vast network in his job search. Hortie wrote from his new job as economic geologist for the Anaconda Copper Company in Butte, Montana, that the head of the Michigan Geological Survey might be leaving.[22] Younger brother Rob Winchell wrote that the director of the Michigan School of Mines at Houghton was leaving, due to politics.[23]

Geology jobs in Michigan, particularly those involving an administrative component, were something Winchell was eminently qualified to hold. He was a Michigan graduate, had cut his geological teeth in the state, had successfully led a complex endeavor for twenty-six years, and had done so frugally and competently. Hortie knew members on the Board of Control at the mining school. The question was, did Winchell want to thrust himself into jobs that might prove more of a headache than a boon? Lucius Hubbard, of the Michigan Survey, informed him that he had quit in exasperation primarily because state authorities had delayed publication of his work for over two years.[24] When Marshman Wadsworth, outgoing head of the Michigan School of Mines (the forerunner of Michigan Technological University at Houghton) and an associate editor of *The American Geologist,* wrote to answer Winchell's queries, he was noticeably reticent about the attractions of the job.[25]

As Winchell pursued the Michigan jobs, his brother Rob wrote about another one in Illinois. A longtime geologist at Northwestern University in Evanston, Oliver Marcy, had died unexpectedly, and the position was open.[26] Northwestern was affiliated with the Methodist–Episcopal Church. Rob was correct: this was a desirable post, and one that would be soon filled by an active Methodist—Winchell's assistant, Uly Grant. He and Avis and their four-year-old son, Addison ("Bump"), moved to Evanston that July and remained at the university thirty-three years.

The political Houghton job became less desirable when Winchell received a letter and newspaper clipping from another associate editor at *The American Geologist,* Michigan university professor Israel Russell, informing him that it had been "practically settled" that the School of Mines at Houghton would be moved to Ann Arbor and folded into the university. The clipping from the *Detroit Tribune* detailed the exorbitant costs entailed by the mining school. Its budget was more than the University of Michigan's, with only one hundred students enrolled. A senator who had visited Houghton declared, "I could not possibly see where the half million dollars it had cost the taxpayers had gone." To compensate for the loss of the school, the article added, Houghton would receive the teachers' college that had been promised to Marquette.[27] None of this materialized.

These job opportunities passed, and volume 4 of the final report was released in 1899 to great acclaim. One recipient of the report wrote thanking Winchell for copies and added, "you have labored long and in a very learned way to advance the interests in both a scientific and economic manner . . . and have added untold thousands of dollars to your state which in time will add millions of money to the people of your state. . . . The state should at once order a new geological survey to work up all the unfinished matter. . . . The cost of your survey $150,000–$200,000 in round numbers is a very small cost for all the labor and facts that have been brought out."[28]

Others echoed this thought. Charles Schuchert wrote, "I was in hopes your Survey would be continued . . . with so much undeveloped territory [Minnesota] cannot afford to be without a geological survey. I supposed your state like others must learn this through experience."[29] And Warren Upham, who had served for so many years on the survey, added, "the interests of Minnesota seem to me to require your continuance in the service of our state as a consulting geologist, for advice and examination of quarried and mining properties, etc. to promote an intelligent development of the resource of Minnesota."[30] Lastly, a Seattle colleague observed: "The loving, self-denying labor you have put into this work, I am sure, has not been appreciated by the state which has benefitted so much by it."[31]

Winchell himself allowed just a glimpse of the great pain of this time when he wrote to John S. Pillsbury, who was president of the board of regents: "I find it a great relief to know that the great struggle of my life is drawing to a close, though misunderstood and condemned by those who ought to be best informed and strongest friends."[32]

In the summer of 1901, a professorship in mineralogy and petrography at the University of Toronto opened up. Seemingly tailored to Winchell's specific skill set, the position attracted widespread interest. Winchell's network bombarded the university with letters of support of him, to the point where a friend told him, "I do not think you need any more letters. A few good ones are better than many."[33]

It was a difficult summer to live through, as Winchell awaited the decision of the university. The family at 120 State Street was nursing ailments. Lottie was so ill that she resigned her long-standing position as secretary of the Minnesota Board of the Woman's Foreign Missionary Society,[34] and Louise suffered serious effects from dysentery.[35] Newton hobbled on another sprain, and what he termed "rheumatism" was painful enough to mention to Hortie in a letter.[36] Adding to the general misery, the weather was unusually hot. From the California interior to the Ohio River valley, the extreme heat excited comment. Minnesota recorded three days in July where the thermometer reached one hundred degrees or greater.[37]

In November came the disheartening news: the Winchells would not be moving to Toronto. The position had gone to Thomas Leonard Walker, a native of Canada with a recent doctorate in the geology and petrography of Ontario's Sudbury Nickel District. He had little postdoctoral job experience, but there was one salient fact: he was thirty-four years old.

"There were a number of very highly qualified applicants among which it was very hard to choose," a friend wrote Winchell.[38] "I may add that the question of age had an important part in the Minister's decision."

Months were passing, and Newton no longer had money coming in, save what he might earn with consulting jobs. Lottie still had boarders residing in some of the eighteen rooms at 120 State Street, and they could rely on rent.

One boarder from the past was promising to become a permanent part of the Winchell clan. His name was Draper Dayton, "Dray" to his friends. He had arrived at the Winchell place in 1897, possibly to attend the university's preparatory academy, for he was only seventeen. He rented the room at the top of the stairs in the old house. His father, George Draper Dayton, a prominent business man in Worthington, Minnesota, had recently bought a business in downtown Minneapolis and maintained an office there, but his mother and younger siblings remained in the family home in Worthington. In May 1898—at the very time Lottie was grappling with Alex's romantic crisis—Draper became very ill, and Lottie moved him downstairs to her room, so she could keep watch over him. Louise, bright and lively and exactly his age, entertained him "most of the time."[39] The next school year, Draper would be off to Princeton, from which he would graduate in 1902. But a flame had been kindled.

Pressed for money, Newton took a freelance geologist job in Butte, Montana, in summer 1903. He was so low on funds that he borrowed from the cash reserves of *The American Geologist* to buy his train ticket out west.[40] Hortie, now based permanently in Butte, loaned him additional funds before his first paycheck. Winchell headed a team of four—two other, young geologists and a cook. Writing Lottie, Newton observed, "Climbing the hills will be the hardest of my work. I am glad that we are to have two horses for riding. I shall use them freely."[41]

The team was based in Bonner, Montana, a logging town with a hydroelectric dam and sawmills owned by Anaconda. But despite the development in town, the geologists were housed in tents. The opening weeks of the field season, in May, were cold and snowy. It had been years since Winchell had spent a summer in a tent, much less in below-freezing weather with two inches of snow on the ground. "I feel a little dispirited this morn. I let the men go out without me. I was not in trim to accompany them," he told Lottie. He added, "I hope I shall be able to keep this going through the summer, but I have misgivings and may have to relinquish the job."[42]

Despite his indomitable spirit, he was aging. Sixty-two this field season, his body ached from arthritis and past injuries. The most worrisome ailment, though, was a delicate one of which to speak: Winchell suffered from that common malady of aging men, an enlarged prostate. In letters to Lottie, he referred to it as "his chief difficulty" or "his trouble," and he kept hoping it would resolve. He persisted, and by late May he was able to write that "I feel stronger than I did two weeks ago. My face and hands are burnt brown and I weigh, I think, at least 10 pounds less than when we began."[43]

While Newton was out in the field, Lottie wrote about yet another mild familial crisis: Louise, now twenty-two, wanted to work as a store clerk, rather than as a schoolteacher, as her siblings, her parents, and grandparents had. And not just any store: Draper's father's new store. In the years since Draper had boarded at 120 State Street, his father had built a six-story building in downtown Minneapolis at Seventh Street and Nicollet and rented it to R. S. Goodfellow's, an established dry goods store. That very year, 1903, he had bought Goodfellow's out and changed the name to Dayton's Dry Goods.

Louise's inclinations might have been a blow to these two fervent schoolteachers, but she was their youngest child, and they had become pragmatic and perhaps a bit worn down over the years. Newton wrote back, "I think you are right as to Louise's going as clerk into the store, tho if she marries Draper she would certainly be in the line of work where he will be for many years probably, and might be qualified to be of use to him, and thus indirectly to herself." Ever the logical thinker, he added, "She ought first to decide that question, and make other things conform to it."[44]

Newton endured most of the field season, despite his "trouble," but in September, as the evenings grew cooler and nights longer, he was back in Butte to see a doctor and attempt the train ride home. He used a catheter two or three times a day and didn't think he could make a continuous trip across the plains.[45]

How humbling to have such a physical limitation to activity that he had formerly gloried in, the rough and stimulating work of field geology. Yet even while he came to terms with this potent sign of decline, he rejoiced in another mark of age: the vigor of his adult children. Lottie had written describing a happy family gathering, and Newton told her, "You may justly be proud of your children. I feel the same way. Our boys and girls are no disgrace to their parents. I am glad Draper was with them [at the party]. I think he will hold up his end, and with Louise will make another shining family star."[46]

Louise would marry Draper, who had just graduated from college, later that year, on her father's birthday.

FOUR YEARS PASSED WITHOUT full-time work. Winchell maintained a pounding pace. He continued as the editor of *The American Geologist,* turning out sixty-five-page issues every month, with Lottie editing. Winchell produced articles, too: "Sketch of the Iron Ores of Minnesota," a review of the initial discoveries, the output, and a conclusion: "There is no theoretical reason to expect that the Mesabi ore is near its exhaustion" (March 1902). He wrote reviews of recent publications in French and German (other editors took on Swedish and Spanish). He opined on articles on topics as diverse as the geology of Labrador, the coral reefs of Formosa (Taiwan), and the evolution of climates (February 1904). Reviewing a book by Frank Taylor on cosmology (March 1904), he observed, "One is tempted to compare this book with the wild and improbable romances of Jules Verne, but such a comparison would be unjust, for Mr. Taylor keeps within the laws of physics."[47]

DURING THIS OUT-OF-WORK INTERLUDE in his early sixties, he also bought a lot across the street from 120 State and built a house. There is no record of how he could afford this. He pitched a proposal to Warren Upham at the Historical Society to edit a ten-volume

"Minnesota Monographs"—a history of the state's beginnings. He proposed to write the first volume, "Geology and Physical Features of Minnesota." Upham liked the idea.[48]

If geological life was not productive, Newton's home life was. He and Lottie welcomed five new grandchildren into the world, bringing the total to seven and including Alex's first. Alex had completed his doctorate in Paris, and he and Clare had moved to Butte, joining Hortie and Ida Belle. Alex became a professor of geology at the Montana School of Mines.

In June 1905, Newton made plans to help Alex with summer work in Portland, Oregon. The city was hosting the Lewis and Clark Exposition, and Alex would oversee the Montana exhibit of rocks and minerals, as he had for the St. Louis World's Fair in 1904. Hortie thought this was a splendid idea, writing Lottie that "the work will not be hard and Portland is a lovely place to spend the summer." But Hortie still worried about his father's health, telling Lottie to "Please call Dr. Roberts"—and he'd pay for it.[49]

It was Alex, though, who offered a radical proposal. He had taken it upon himself to write Dr. Charles Mayo in Rochester, Minnesota, when he had learned that enlarged prostate glands were treatable with surgery. Mayo wrote back that operations to remove the prostate gland were now "very safe. Considering the age and infirmities of these patients the mortality is low—not exceeding 4 or 5% . . . over 90% are practically cured."[50]

Alex also wondered about cost. Mayo replied that they didn't have a set fee "but make our charges something within the means of patients." For postoperative costs, "rates were $7 per week in the wards and from $12 upwards in private rooms."[51] Stays of two to three weeks were usual. He could bring his father there anytime.

Alex told his father, "Now it seems to me that this is troubling you so much as to make an operation advisable . . . of course, also, I shall not *urge* it, as it is too serious a thing." He also wrote, "I want to say emphatically that if you decide to have the operation it would be much better to have it immediately—before you go to Portland—and so have good health while there, than let it drag on, getting worse probably all the time, and all the time oppressing you."

Lest his parents, who had so little income at this point, agonize over money, Alex added, "the matter of cost should not be considered at all. I shall be glad to pay all the costs if you have the work done." It would not be a loan, either: "I would be only repaying you a small part of what you have given me." Lastly, he added—and one might imagine a twinkle in his eye when he wrote it—"I don't think Dr. Mayo will talk about 'infirmities' in your case after he knows you."[52]

There it was. An entirely new idea, requiring courage on Newton's part and trust in Charles Mayo, for abdominal surgery in the time before antibiotics ran a grave risk of fatal infection. Winchell kept the missive and penciled in a commentary: "A model son's letter." Then he and Lottie had to make a decision.

19

THE ARCHAEOLOGIST

1905–11

Newton Winchell shifts in his hospital bed, glances at the crucifix on the bare wall, then turns his gaze to the window. An April day outside. Blue skies, red flowers on the maples tossing in the breeze, a robin singing. He has been here three weeks, and although he feels stronger now after the surgery, he is not well enough to go home, and he knows it. The ordeal shocked his system; he nearly didn't make it. He is lucky to be alive.

Inside, the hospital halls are quiet. The nuns speak in whispers. Noise jars the delicate process of healing. Doctor Mayo comes every day, Nurse Lynch is always in attendance, but time creeps along here. At least he has the newspaper. The Rochester Post and Record *is his daily entertainment.*

IN MARCH 1905, WINCHELL UNDERWENT a prostatectomy under the skill of Charles Mayo. It is a difficult surgery because of the position of the gland and its relationship to the bladder. Its removal was attempted by very few surgeons in 1905. Dr. Mayo was one of the best.

The practice of the Mayo brothers, Charles and Will, was expanding rapidly in 1905. Surgeons and surgical consultants, they had abandoned a general practice to their partners,[1] and in 1905, they would perform over two thousand abdominal operations, an average of almost six per day. Because they were early adopters of antiseptic practice, the surgical mortality rate was very low, attracting even more patients to their clinic.

When he was well enough to travel, Newton joined Alex and his family in Portland, Oregon, to recuperate in the mild Pacific Northwest. He took his camera and spent time perusing the geological displays at the Lewis and Clark Exposition held in Portland that summer. He was particularly taken with the Willamette Meteorite, calling it "the most wonderful single object in the mining building."[2] The meteorite was big—at sixteen tons, it was in 1905, and remains today, the largest meteorite to have been found in the United States. Unlike the meteorites Winchell had collected for the university museum, there was no record of its fall, but the Native Americans of the Willamette Valley knew of it and called it "tomonowos," "visitor from the sky."[3] There was buzz that after the exposition, the monstrous rock, shaped like an inverted Liberty Bell, would leave the Northwest and be acquired by an East Coast museum. Winchell hoped that would not happen,[4] but money speaks. The Willamette meteorite has been on display at the American Museum of Natural History in New York City since 1906.

During his summer in Portland, Winchell also addressed another nagging concern: the fate of *The American Geologist.* In 1905, the journal was in its eighteenth year. Winchell, who had been the chief editor for all those years was sixty-five; Lottie, working unpaid the entire time, would be seventy on her next birthday. The *Geologist* had become, as Hortie observed, "burdensome to my Father and Mother in their declining years."[5] A new journal with an emphasis on mining, *Economic Geology,* had launched, and its owners offered to buy the *Geologist*'s subscription list for $1,000.[6] Indeed, many journals now specialized in various geological fields and directly competed with the *Geologist,* which had once been the only venue for publication.

In June 1905, Winchell sent a circular letter to all twelve associate editors, seeking input on the journal's fate. He floated the idea of someone else taking over as editor in chief, but most replies were in the vein of Oliver Hay's of the American Museum of Natural History: "I am not in a situation to assume the responsibility of managing such a publication. . . . I really trust that you may find your way clear to continue."[7]

But there was more to the situation than an aging editor. Uly Grant, in the field in Valdez, Alaska, observed that "somehow, the best papers do not seem to be coming to the 'Geologist.'" He advised his father-in-law to "accept the offer . . . and free yourself of a big task and a big responsibility."[8] Florence Bascom of Bryn Mawr, the journal's only female editor, was more blunt: "I should myself rather see the Geologist terminate its career at once in a dignified fashion than that it should meet with a slow death."[9]

Hortie brokered a deal with *Economic Geology:* $1,000 for the subscription list of more than three hundred names; the name of *The American Geologist* would appear in the title of the new journal for a period of time, to make it an actual successor and not a takeover of the older publication; and Hortie would be the middleman, buying out the owners of the *Geologist* and receiving remuneration from *Economic Geology.*[10]

Newton was not thrilled with the arrangement. In fact, Hortie termed him "rather bitter." Perhaps the elder Winchell had hoped a younger geologist would step forward to continue his journal. Hortie told his father that he did not "wish to force the plan if you don't feel that it is a relief to yourself."[11] If the plan was carried out, though, it must be done in an upbeat manner to benefit both journals. In a subsequent letter, he warned, "the coming consolidation of the American Geologist and Economic Geology should be made in a tone of cheerfulness and progress. . . . there should be nothing resembling an obituary notice in this announcement."[12]

Winchell heeded the advice of his elder son. A lengthy editorial comment in the November 1905 issue of *The American Geologist* announced the coming merger. This issue, it said, would be the last. Winchell assured his readers that although the *Geologist* had operated in the red initially, it had been self-sustaining for the past ten years. It had, in fact, run a modest surplus for the past five to six years. The *Geologist* had never been in better financial shape. "The sole reason," Winchell wrote, "for surrendering this charge is the desire on the part of the managing editor, with advancing years, to find time for some other contemplated work."[13]

He ended his final issue with three book reviews written by him, of a book on the origin of granites, of one on petrographical studies of Colombian rock, and of one translated from the French. He gave Hortie the final review, of the book *Economic Geology of the United States* by Heinrich Ries.[14]

He had closed another chapter of his life.

As Winchell shed the last of his geological responsibilities, another door opened. In February 1906, the Minnesota Historical Society offered him a position in archaeology, curating its collections and writing a history of the Native Americans in Minnesota.[15] The association between Winchell and the society was a long one, and he had served on its governing board in 1904. Undoubtedly, the presence of Warren Upham, with whom he had retained a warm personal and professional friendship, also smoothed the way for a job offer. Upham had held the paid position of society secretary since 1895.

The position opened when Jacob Brower died unexpectedly in June 1905. Brower had worn many hats in early Minnesota history. A surveyor by training, he had been tasked in 1889 with mapping the Lake Itasca region and determining which small creek trickling into the big lake could be considered the "true" headwaters of the Mississippi River. He had been appointed park commissioner in 1891 when Itasca became the first state park. Brower spoke forcefully for the park when lumber barons, who retained logging rights in the park, threatened the pines. In May 1903 he scrawled in his field notebook that he was going with a statesman to Itasca to view the situation, with the intention to "save Itasca State Park from the despoliation of lumbermen."[16]

In these years, he also raged over the plight of the Mille Lacs Band of Ojibwe, who had been cheated of their land and reduced to abject poverty and were being harassed into leaving their ancestral land for reservation life at White Earth. Brower was incandescent and offered his services as a go-between with the federal government and band members.[17]

Brower was white-haired and over sixty; overworked and overstressed, he himself thought he might not live long enough to complete his archaeological research.[18] After his death, it was Newton Winchell who stepped in to inherit Brower's field notebooks, maps of mounds, and thousands of artifacts: stone hatchets, spear tips, arrow points—all dished up with Brower's passion.

The job offer was welcome in many ways. It was necessary for income. Pensions were not common in 1900, and there had not been one with the survey. But it also allowed the sixty-six-year-old geologist to pursue a long-held interest. When, in his *American Geologist* swan song, he had referred to finding time "for other contemplated work," it was this he meant—a concentrated plunge into the field of archaeology.

In 1906, Winchell's new employer, the Minnesota Historical Society, was already a fixture in the state. Its library was a busy place, frequented by so many visitors that often there were not enough tables and chairs to seat them all in the reading room.[19] Its forty-three thousand bound volumes eclipsed that of other states.[20] It owned issues of nearly every newspaper ever published in the state and an equally large portrait collection. These precious documents of state history had moved to a new capitol, designed by Cass Gilbert, in 1905. The society was relieved to get them into a fireproof building, for it had lost material when Minnesota's first capitol burned in 1881. Moreover, recent conflagrations served additional warning. Fires had consumed Wisconsin's state capitol and the beloved main building at the University of Minnesota only the previous year.

But Winchell worked in the former capitol down the hill from the new capitol, at the corner of Wabasha and Exchange.[21] In his "encore career," Winchell now commuted from Minneapolis to St. Paul via streetcar. Lines ran down Washington Avenue and along University Avenue, depositing him in front of the old capitol at Tenth and Wabasha. The streetcars ran every five minutes and took about thirty minutes for the trip. But because each city required its own fare, Winchell paid double for crossing that city line—five cents getting on and five cents getting off—each way.[22] The society paid him adequately but not extravagantly: $1,800 per year, about $46,000 in today's terms.

The old capitol had housed the society since 1883. Tucked away in basement rooms, which nonetheless were lit by large windows, Winchell presided over the Department of Archeology, with an office and a laboratory from which he examined, described, and cataloged artifacts. There was a warren of offices in the basement, housing a variety of civic organizations, and there was a fireproof vault, in which were stored the most valuable items, including Brower's journals. But the museum, with its display cases of Native American artifacts, was in the basement of the east wing of the new capitol, so Winchell frequently made the uphill trek to the splendid white edifice.

The new job put Winchell in the heart of a city in a way he hadn't been at the university. St. Paul bustled, streetcars clanging, horse and carts mixing with newfangled horseless carriages on the bricked streets. On a high hill above the business district, Bishop John Ireland's St. Paul Cathedral was under construction. On Capitol Hill to the east of the new capitol, a Richardsonian Romanesque mansion, built by former governor John Merriam, dominated a wealthy neighborhood. It would later house the St. Paul Science Museum.[23] Immense red brick and sandstone hotels and office buildings cast shadows in the morning and afternoons across the busy streets. Across Cedar Street from the old capitol, the bell of Central Presbyterian announced services.

Winchell rolled up his sleeves and tackled this new field. Always an autodidact, he read widely and studied Brower's vast array of artifacts. In 1902, he had taken a trip to Lansing, Kansas, with Brower and Upham to inspect human remains discovered in glacial deposits, and it had rekindled his interest in the geological aspects of archaeology. Later, he would become a pioneer in applying his expertise.[24] He had been interested in Native Americans for most of his life. His geological field experience had put him in contact with Native people and artifacts in both Michigan and Minnesota.

He had described mound construction whenever he encountered them on survey trips, calling them "earthen works,"[25] and Brower had left him beautiful, hand-drawn maps of a multitude of mounds. Brower had also identified contacts in various areas with an abundance of mounds, and Winchell wrote to them. The mounds were all over Minnesota, in Red Wing and the Lake Pepin area, near International Falls, at the mouth of the Bowstring River, and at the mouths of both the Big Fork and Little Fork Rivers. There were remarkable mounds up and down the broad Mississippi River valley, and Winchell's contact at Trempealeau, Wisconsin, George Squier, had begun careful excavation of them in consultation with the Wisconsin Archaeological Society.[26]

The mounds were often disrupted by those who plundered without regard to the humanity of Native peoples and without the reverence due a burial site.[27] Typically, the mounds were of significant age. That turned out to be the case with the Trempealeau mounds, which proved to be four thousand years old.[28]

People also notified Winchell of unusual mounds near Mankato in Cambria Township of Blue Earth County. Sixteen large mounds (forty feet in diameter and twelve feet high) overlooked the Minnesota River in a picturesque spot. They contained artifacts as well as human bones.[29]

Pressed for time compiling Brower's note and artifacts, Winchell delayed a trip to Cambria, and Mankato lawyer Thomas Hughes wrote again. They had found "pottery . . . all well made and burned and quite artistically decorated . . . also pieces of bone or animals, bone needles, flint chips and flint arrowhead and spearhead, whetstones, clam shells. The depth at which the pottery is found indicates a great lapse of time. . . . wish you could come and investigate the site yourself as you are much better qualified to judge of such matters than we are who have no experience." Hughes added (at which Winchell may have ground his teeth): "Quite a number of Mankato people had been up in Cambria to investigate the find and all come home loaded with pieces of the pottery. . . . if you come down, let me know and we will arrange to take you up in an auto from Mankato."[30]

Winchell did eventually get to Cambria and hired William Baker Nickerson, a slender, spectacled man with a walrus mustache to excavate the site. Nickerson was an amateur in a field that was transitioning from amateurs to professionals, but he was highly intelligent and meticulous, characteristics that attracted Winchell. Archaeology seldom paid well, and Nickerson worked on the railroads to earn his keep. He preferred the night shift, so he could excavate promising finds plowed up by railroad construction during the day.[31] He was slow to arrive at Cambria, as he was just finishing up a site in Iowa, but within days of investigating the site, he wrote his boss, "I am convinced that we have uncovered evidence of human occupation of this site very much older than anything we had anticipated," and he took the initiative to call for a photographer from Mankato to record the bones in place in the sand, with the undisturbed strata over them. Then he told Winchell that he would leave the dig open until the boss could get down there.[32]

The settlement at Cambria *was* old. Archaeologists have dated it as 1050–1200 CE, a Late Prehistoric settlement and part of the Plains Village culture of the prairie–forest border in Minnesota.[33] The pottery ornamentation had ties to those of Cahokia, the great culture centered on the midcontinental Mississippi River now known to date from 400 to 1400 CE.[34]

Winchell did a lot of fact-checking on Brower's voluminous notes while he organized and began writing his manuscript. He had a question about the Dakota chief Wabasha (variously spelled Wapesas or Wapasha). The hereditary line was confusing. He queried Joseph Buisson, age ninety-four, an old settler of the town, who arrived in the area in 1838. Possessing a remarkable memory, Buisson had known three generations of Wabashas, knew where they had lived, and knew the Dakota habits of seasonal migration. He pointed out to Winchell that the city named after the chief *should* have been spelled "Wapesa," which is closer to the Dakota pronunciation. It means "Red Leaf."[35]

One of Winchell's concerns was the proper identification of Indian portraits and the correct spelling of their names. This was no easy matter, for the names written in English were long and phonetic: for example, Kah-bi-nung-gwi-wain of Lake Winne-bay-go-shish, known for his caustic wit and use of sarcastic epigrams, and Bud-ig-oonce, "a man of remarkable intelligence and influence."[36] After several drafts, additions, and corrections, Winchell's source pronounced the spellings "sufficiently good."

Winchell's treatise on Minnesota's Native peoples drew not only on Jacob Brower's collection and journals but also on those of two earlier individuals, Alfred Hill of St. Paul and Theodore H. Lewis. Hill was an educated Englishman who had arrived in St. Paul in 1855, intrigued by the numerous mounds existing, sphynx-like, on the banks of the Mississippi River and elsewhere. Learned and in later years reclusive, Hill amassed an impressive collection of artifacts. When the mounds came under increasing destruction with post–Civil War development, he hired T. H. Lewis, an Ohio surveyor, to methodically map those that remained.[37] Lewis was a fortuitous choice; his technical competence is admired to this day.[38]

Immediately upon Brower's death, the Minnesota Historical Society had purchased Hill's archaeological reports, surveys, maps, and field notes before they went to anyone else. Winchell, then, drew on all three early collections in writing his treatise. Since Brower and Lewis had maintained a mutual animosity, the end result was a more comprehensive report than would have resulted had Brower lived to write up his own work.

IN THE MIDST OF FACT-CHECKING DATES, spellings, locations, and other minutiae, Winchell waded into investigation of an artifact of a different kind: a large slab of graywacke, uncovered in west-central

Minnesota a decade before, on which had been chiseled mysterious symbols.

Known as the Kensington Rune Stone, the artifact had been unearthed by a Swedish immigrant farmer who had been grubbing out a field on his farm near Kensington, Minnesota, in 1898. The roots of an aspen sapling about ten to seventy years old entwined the stone, suggesting the stone had been in place before the tree took root. On one face of the tombstone-like slab were chiseled what seemed to be ancient characters and were later identified as Scandinavian runes. The inscription was translated by academics:

> 8 Swedes and 22 Norwegians on an exploration journey from Vinland westward. We had our camp by 2 rocky islets one day's journey north of this stone. We were out fishing one day. When we came home we found 10 men red with blood and dead. AVM save us from evil. We have 10 men by the sea to look after our ships, 14 days' journey from this island. Year 1362.[39]

In the first year after the rune stone's discovery, it was translated by a University of Minnesota professor of Scandinavian languages and pronounced a hoax. The matter rested for almost a decade until it was resurrected by a Norwegian American scholar from Wisconsin doing research in Douglas County, where the stone had been found, who visited the farmer and somehow procured the stone—whether given it or got it on loan was, and still is, unclear.

This scholar, Hjalmar Holand, published his book on Norwegian settlements in 1908 and devoted a chapter to the rune stone. Holand was enthusiastic and emphatic on the authenticity of the stone and defended it on several different levels.[40] In July 1908, Winchell wrote to him, asking for a translation of the inscription and asking a question that no one had posed before: how might this stone have interacted with Glacial Lake Agassiz, whose highest level, marked by a beach line, the Herman beach, was in the vicinity?[41] Winchell apparently was unaware of the date on the stone.

Holand replied within a few days, supplying Winchell with his own translation and assuring him, "The authenticity of this inscription (which long study has convinced me is genuine without doubt) does not depend upon the date of the retirement of Lake Agassiz."[42]

Months passed. Winchell returned to pondering Dakota and Ojibwe culture. The Norwegian Society of Minneapolis formed a committee to investigate the stone that included a prominent local physician and surgeon, Knut Hoegh. Dr. Hoegh made the trip to Kensington in summer 1909, interviewed the farmer, whom he deemed "trustworthy" but "unschooled"[43]—the farmer had a shaky grasp of English—and he reviewed arguments supporting the hoax theory. His report suggested he believed in its authenticity.[44]

The rune stone's significance could not be overstated. If indeed authentic, it would be the earliest document of Europeans in North America. Although the date, 1362, fell later than the Viking period, Holand was quick to point out there were records of a Norwegian expedition to Greenland in 1354, which he asserted returned to Norway in 1363, and so lent plausibility to the date.[45]

Winchell made plans to travel to Kensington to investigate the stone through interviews and site visits. His first impulses, to see the site firsthand and to seek expert opinion, fit with his lifelong approach to scientific inquiries. Did alarm bells go off in his head when Holand wanted thirty-five dollars to write up a report on the stone for the Historical Society, instead of submitting a paper to a peer-reviewed journal? Or when he stated that the stone was "genuine beyond doubt" before the stone had been studied by experts?[46]

Winchell returned from Kensington with geological details of the site. The slab had been found on a hill surrounded by a marsh, which in previous centuries would probably have been a lake, hence supportive of the inscription's mention of "this island." He had looked around for similar boulders and found ones of graywacke exactly like the stone. He had interviewed the farmer and deemed him honest. In limited English, the farmer told him that Swedish and Norwegian children learned about runes in school. Other interviews done later by Winchell and others produced a wealth of speculation on who had the ability to perpetrate a clever fraud, who might have been in on it, and what might have been a motive.[47] Winchell also asked the farmer how the rune stone had passed out of his possession. The farmer emphatically claimed that he had never sold the stone to Holand, who nonetheless now had it.[48]

The Historical Society read Winchell's initial report and contracted with Holand to put the stone on display in the new capitol, "where it may be freely inspected by visitors."[49] Three weeks later, upon inquiry, Holand offered to sell the rune stone to the society for the fabulous sum of $5,000.[50] Before the stone was delivered, though, Holand took the big, heavy slab to Chicago to be examined by experts there. The linguists, one and all, considered the inscription fraudulent for various reasons. Dr. George Curme of Northwestern University thought the stone's use of umlauts, which are modern, was enough to be decisive.[51] C. N. Gould of the University of Chicago wrote Warren Upham a twenty-two-page typewritten analysis citing linguistic inconsistencies, including a word chiseled on the stone that wasn't in use until three hundred years later.[52] Gould, who had been at Holand's lecture in Chicago, went further and sought to ascribe psychological reasons for the Norwegian American's defense of the stone, observing that "he was untrammeled by facts," and writing that "[Holand] is the main part of the Kensington stone. But for him it would have remained before the farmer's granary door."[53]

Winchell, who was inclined to think the rune stone authentic on geologic and geographic grounds, had counterarguments for all of Gould's claims. Without any linguistic training, he noted flaws in Gould's points. For example, that one word that had not appeared in literature until the 1600s? It could have been in vernacular usage much earlier.

It might seem sheer hubris that Winchell would buck the consensus of linguists, but it was in keeping with his personality. He had great confidence in his mental acuity and had run counter to prevailing opinion many times in his long career. He pointed out that for linguists, the linguistic evidence was paramount, and other evidence—geological in nature, say—was pertinent only if linguistic evidence affirmed the stone.[54] Newton Winchell considered geology primary to everything.

Winchell's report of the Museum Committee to the society ran seventy-six typewritten pages and was delivered in April 1910. The Museum Committee went on record "rendering a favorable opinion of the authenticity of the Kensington Rune Stone"—if the reasoning of the report could be verified by a competent specialist.[55] But as the summer wore on, the committee, and particularly Upham, were gnawed by doubts. Another linguist had severely criticized their April report, and in July 1910, the committee softened its language in support of the stone.[56] In the end, it never paid Holand for the rune stone, which it deemed questionably his. Holand took the stone to Europe in 1911, where it was put on display in Norway and France. Experts in Norway judged it to be fraudulent.[57]

One odd argument used in support of the veracity of the stone, which Winchell labeled as "collateral evidence," was the belief that there was European blood in the Mandan Indians of the North Dakota plains. "All travelers who visited them reported instances of light-colored hair and skin and blue eyes," Winchell observed in his April 1910 report.[58] Including this piece of "collateral evidence" was a stunning breach of Winchell's cardinal rules of observation with his own eyes (he did not travel to the Mandans), of not relying on hearsay, and of not venturing into areas (genetics) of which he was not informed. He even wrote up his thoughts and submitted them to the Smithsonian's *Anthropologist* magazine for publication. The paper, "The Mandans and the Norseman," was tersely rejected by the magazine's board, which "on various reasons, express[ed] themselves adversely to its publication."[59]

Perhaps the robust rejection took the wind out of Winchell's sails. The matter of the Kensington Rune Stone soon quieted for lack of conclusive evidence, and he returned to the task for which he had been hired, writing a definitive account of Minnesota's Native Americans. The story of the Kensington Rune Stone was not over, however. Because no argument was ever rock solid, because no one ever claimed to have perpetrated the hoax, and because certain claims in the stone remained unexplained, it periodically rises to scrutiny, even today. Some people just want the story to be true. It seems that Winchell was one of them.

THROUGHOUT HIS LIFE, Winchell had tended multiple irons in the fire. In the years in which he worked on the Kensington Rune Stone, he turned seventy, released a textbook based on his Paris work, *Optical Mineralogy,* coauthored by his son Alexander—a textbook still in use today—and oversaw the construction of a new house at 501 Union Street at East River Road. The first in the neighborhood, the house nestled into the hillside and overlooked the Mississippi River. Winchell and Lottie christened it "Bella Vista" and moved in July 1909.

Later that summer, Hortie and Alexander were together on the S.S. *Ohio,* heading to Alaska to join Uly in the field, when it hit an uncharted rock (now named Ohio rock) and sank.[60] The brothers scrambled into a lifeboat before the ship went under. The family that gathered for Thanksgiving dinner in the new house that year with all but Avis and Uly attending had much to be thankful for.

The Aborigines of Minnesota was released in summer 1911. A whopping 743 pages, it was illustrated with 36 halftone page plates, 26 folded inserts, and 642 figures inserted in the text. It was an exhaustive report covering human presence in the area now called "Minnesota." Winchell described the entries of both the Dakota and Ojibwe peoples into the state and what tribes they displaced, dating the events when possible. He related origin myths, food sources, utensils and weapons, and means of transportation. There were maps from the 1600s, the earliest drawn, depicting who lived where. There was a list of Ojibwe personal names that detailed dates, information about the person, and included men and women. This list ran for twenty-five pages, an estimated 1,400 names. Winchell wrote in the Introduction that he aimed to make the work useful for the Ojibwe as well as all scholars. The Dakota were not mentioned as an audience, presumably because the tribe had been largely removed from Minnesota following the Dakota War of 1862.

Of that traumatic event that punctuated Minnesota history for all time, Winchell, a man of his time, had this to say (one can almost hear him sigh and gaze off into middle space): "Perhaps the most comfortable view that can be taken of the deplorable events that marked the termination of the rule of the Dakota in Minnesota, is that which is also the most philosophical . . . [in] the history of the world . . . might, right or wrong, has prevailed."[61]

The Aborigines of Minnesota garnered accolades for Winchell. How could it have been otherwise? It was Winchell at his best: comprehensive in scope, methodical in approach, meticulous in detail, a worthy companion to the voluminous *Geology of Minnesota.*

Archaeologist Warren Moorehead, who would later be acclaimed for his work on the Hopewell culture of Ohio, considered it "the most important work yet published on that region."[62] Historian William Watts Folwell, the man who had brought Winchell to Minnesota in 1872, was very pleased with the result. "It was a kind of good fortune for the state," he wrote, "that you were induced to take hold of the material and put it into useful form."[63]

Within weeks of the book's release, Warren Upham sent Winchell a letter notifying him that since the task for which he had been hired was complete, his agreement with the Historical Society was over.[64] But Winchell, now seventy-one, had another research area he wanted to pursue: the question of how long humans had been in North America. He had been able to keep abreast of the debate in the archaeological community while he worked on *The Aborigines of Minnesota* and the rune stone, and now he was eager to throw his undivided energies into the question. He persuaded the society that his endeavor would bring a luster to it, even though much of the fieldwork would fall beyond Minnesota's borders. They agreed to keep him on.

20

WINCHELL THE ROCK

1911–14

NEWTON WINCHELL HAD LONG BEEN interested in the possibility of early human presence in North America. In 1877, when the survey searched for evidence of Zebulon Pike's stockade at Little Falls, his eye had been drawn to pieces of chipped white quartz scattered where old-timers had told him the fort might be. The chips looked unnatural, as if they were of human workmanship. After gathering some, Winchell found an arrowhead, as well. The chips seemed localized to that spot, but later Winchell also found them by excavating deep within a riverbank.

There were mounds at Little Falls, and at first, Winchell thought that the people who had built the mounds had also made the chips. But the mounds were constructed of sand, which had been spread in a layer by the postglacial Mississippi. The chips were older, buried four feet deep in the sand layer, dating to when the river was wide and draining a melting glacier, laying down sand it had picked up. There were so many chips, Winchell suggested that spot might have been "a grand manufactory in the neighborhood" where toolmakers turned out implements and weapon points.[1]

In 1902, he revisited the site with Jacob Brower and amended that conclusion. They proposed that the quartz-workers (as he called them) had produced the chips about 1,000 to 2,000 years after the retreat of the ice margin from the area, or about 6,000 years ago, a time when the northern half of the state was still encumbered by ice. He thought they might have been the predecessors of the "Eskimos."[2]

Archaeologists of the time broadly categorized early humans into "Paleoliths" (the first part of the Stone Age) and "Neoliths" (the New Stone Age, immediately preceding the development of written language). Paleolithic people were thought to have crafted crude, rough tools made of stones and bones; they were hunters and gatherers and lived in small groups. Neolithic people were distinguished by the presence of ground and polished stone tools, and others crafted of copper. They cultivated some plants and animals and did not have to be constantly on the move in search of food. Today we understand the contrasts to be more complex, particularly in light of the exquisite workmanship of some North American artifacts.

European archaeologists had made this distinction decades before Winchell's time. The caves in southern France, in the Dordogne, had been discovered in 1868 and yielded skeletons that later were dated to 28,000 years old. Workers in South Wales in the United Kingdom had uncovered a human skeleton with numerous tools that was later dated to 33,000 years.

The question of whether people lived in North America during the Ice Age was raging among American archaeologists. The year after Winchell had described the quartz chips in the survey's sixth annual report (1878), a Little Falls schoolteacher, Frances Babbitt, had taken up the inquiry. She located other rich deposits of chips, subsequently publishing and presenting her findings at the 1883 AAAS meeting in Minneapolis.[3] The Little Falls quartzes, as they were called, were bandied about as an example of Paleolithic artifacts.

A rare woman in archaeology, Babbitt was in her midfifties when she began her work on the Little Falls quartzes. During her research, Warren Upham, the glacier expert, had helped in determining the geologic time period of the chips. By 1902, when Winchell had rekindled his interest in the topic, Babbitt had died, but the debate had not.

Winchell's reawakening to the question of people of the Paleolithic came in the heat of summer 1902, on a terrace of the Missouri River valley. That year, farmers in northeastern Kansas near Lansing had come upon human remains while digging a root cellar. Jacob Brower thought Winchell and Upham might have some interest in these bones found deep in twenty-foot-deep windblown silt, called loess. They *were* interested and took a train to Lansing, joined by geologists from the University of Kansas. They examined the material atop the remains and later dated the skeleton to about 12,500 years old.[4]

Both Winchell and Upham were conversant with Paleolithic artifacts from Europe. Tools in the Somme valley in northern France were 100,000 years old. Humans had lived that long ago in Europe, so it was entirely possible that the "Lansing man" was 12,500 years old. That age made it the oldest North American human skeleton to date.[5] The individual associated with the skeleton might have coexisted, Upham thought, with extinct species of bison, moose, camels, llamas, mastodons, and mammoths. In fact, some of these may have gone extinct because of the skill of the Paleolithic hunter.[6] Upham was one of the first to propose the idea, which even now is a topic of contention.[7]

In 1902, Winchell was at the helm of the Geological Society of America. As president, he addressed the society on the topic at its December meeting. It may have been an introduction to the idea for some; for Winchell, the speech served as a bridge from his former to his future life. In the address he discussed the find of a human skeleton in glacial deposits, and, in true Winchellian fashion, methodically, meticulously laid the case for the skeleton's great age. He noted that in the 1870s, geologists (including himself as a young geologist in Minnesota) first entertained the new idea that there had been two glacial events, with an interglacial period long enough for big trees to grow. Now, Sam Calvin, state geologist of Iowa, whom he credited with having done the greatest work in parsing the Pleistocene, detected five glacial epochs, separated by four interglacial epochs. This meant that there had been a very long time since the first onset of ice.[8]

The great river valleys of the interior of North America—the Mississippi and the Missouri—are marked by a silty material, loess, that was carried into the valleys by meltwater and then blown by the wind. (Charles Lyell, author of the world-changing *Principles of Geology,* had proposed this, and Winchell agreed.) The loess's ultimate source is glacial till, taken up by water from the melting glacier and spread broadly over a plain. The Missouri River basin, in which the Lansing skeleton was found, was not reached by the most recent glaciation, the Wisconsin; therefore, its loess was a product of an earlier ice period. This marks any artifact in it as much older than 8,000 years, Winchell's own estimate of the retreat of the last ice mass.

Another highly respected geologist, T. C. Chamberlin, also visited Lansing and disagreed. The tension between two interpretations is a mark of a topic of active research,[9] and Winchell relished good, meaty disagreement among geologists. Later, he would encounter hostility from archaeologists who didn't believe geology had much to contribute to their field, and then the conversation assumed overtones of jockeying for intellectual supremacy.

Winchell's interest in the artifacts of Paleolithic people in America had to be shelved to write *The Aborigines of Minnesota*. It was hard to do, because Brower had collected many artifacts from an early site in central Kansas, which begged to be described and analyzed. But once *The Aborigines* was published, Winchell had a path clear to pursue this question that had occupied a corner of his mind for decades.

Winchell's real contribution to the field would be his geological knowledge, and in this he was aided by James Todd, a geology professor at the University of Kansas at Lawrence. Winchell had known Todd for years. He had, in fact, once hired him for the geological survey.

Todd's research was in the Pleistocene of northeastern Kansas. He supplied Winchell with information on the terraces of the Missouri River, drainage levels from the retreat of the Kansan glacier, and other aspects of glacial geology that could help Winchell explain and date the mallets, hoes, arrowheads, and spear tips in Brower's collection. In May 1912, Winchell and Todd spent two weeks in northeastern Kansas, examining the terraces and what remained of the terminal moraine of the Kansan glaciation.

A stiff wind ruffles the top of the tallgrass prairie, sending out waves like a sea, as Winchell and Todd head out into the May morning. Though the grass beneath their feet is fresh and green, the rising sun is already hot. There's a reason the two aging geologists opted to examine the Missouri's terraces in May and not wait for summer.

Todd halts in his path and gestures to the hills in the distance, capped with chert gravels. These, he remarks, are preglacial, were not disturbed by the first mass of ice edging down from the north. Were not disturbed at all. The landscape is nearly devoid of trees, save for cottonwoods clinging to the streams as they snake through the hollows. The two squint, imagine an ice sheet melting on the margins. Imagine a shimmering lake formed of meltwater yawning from ridge to ridge, shifting the fine particle runoff and smoothing it level.

A meadowlark's liquid song hangs on the breeze. Their jackets flap, they push their hats down tight so the crown is snug, and continue on. How bracing to be out in the open with magnificent country all around. How much better to be at work under the limitless blue sky than to be confined to a classroom, endlessly correcting exercises.

WINCHELL SPENT THE NEXT TWO WEEKS visiting numerous museums in Kansas, Missouri, Ohio (including a promising site in Hopewell County), and Wisconsin. He was searching for artifacts with a certain gloss to their surface that he recognized as signifying aged chert—all the artifacts in this location were of chert.

Winchell had held many a rock in his hand, and these artifacts were rocks. He knew that chert changes color when exposed to air, going from a bluish color to yellow. He had observed that the oldest implements in Europe had a shiny surface, a gloss. He thought he could detect different sets of chip marks on the stone, older marks and younger ones. He was confident that he could ascertain the age of them, relative to the ice age in Kansas.

Before he published an account of the artifacts, Winchell sent the manuscript to Frederick Hodge of the Bureau of American Ethnology in Washington, D.C., where Hodge was ethnologist-in-charge. Hodge was thoroughly versed in archaeology but had no background or interest in geology, and little appreciation for what a knowledge of rocks could contribute to analysis of stone tools, and he was not impressed. While "it [was] very difficult . . . to express an opinion of your conclusions without the evidences on which they are based," he didn't believe that artificial "glossing" was an indication of great age—why couldn't it be due to weathering or to chemical action over a few hundred years?[10]

Winchell fired off a characteristic reply. He addressed each of Hodge's objections patiently and in great detail. Concerning the gloss, he said that he had examined some of the European flints that had a similar patina. "You will see that nothing short of several thousand years would produce such a gloss." As to weathering or chemical action? "Chert is almost immune from attack by the weather or chemical action from acids contained in the soil or in the air." He concluded, "Recent artifacts are never glossed like these."[11]

In 1913, Winchell published *The Weathering of Aboriginal Stone Artifacts* under the auspices of the Minnesota Historical Society. He called it his "little book," and at 186 pages, compared to *Geology of Minnesota* and *Aborigines,* it was.

In the preface, he noted that the antiquity of man is a most important question, and stated that the leading American authorities were sixty-six years behind the Europeans in acknowledging that human beings lived in North America in preglacial times. He pointed the finger at "powerful influences that are localized in Washington,"[12] namely, Frederick Hodge and his Bureau of American Ethnology, or BAE.[13]

The BAE valued field experience over academic training and had little contact with European archaeologists or artifacts. Ironically, it was dismissive of amateur collectors, whom it felt had not mastered the discipline. It had been critical of Frances Babbitt's work and now attacked Winchell as well.[14]

Some established archaeologists were not dismissive of Winchell's geological contributions. Frederic Putnam of the Peabody Museum of Archaeology and Ethnology at Harvard University was broadly trained in natural history (Louis Agassiz had been an early influence) as well as archaeology, and was open to good scientific work. Putnam was known to mentor amateur archaeologists from afar.[15] One was Babbitt, who had written to Putnam about the Little Falls quartzes, and another was William Nickerson, who in 1913 was at work for Winchell at Cambria. Putnam was in contact with several tireless archaeologists, Charles Abbott and Ernest Volk, who had worked for years on artifacts found in till in the Delaware River valley near Trenton, New Jersey. Abbott had felt the full extent of BAE criticism.[16] Winchell was especially intrigued by and sympathetic toward Abbott's efforts, and the two conducted a warm and lively correspondence throughout.

"You ask about criticism on my little Kansas book," Winchell wrote in the fall after its release. "All who have expressed an opinion commend it, and especially the *method.* [Warren K.] Moorehead says I have opened up a new domain of research to American archeology."[17] Moorehead was a wonderful advocate. Connected with Phillips Academy in Andover, Massachusetts, in 1913 he, too, had been taken under Frederic Putnam's wing in his early career, and while working for him had discovered the Hopewell culture "type site," which was yielding information of very old human habitation. Moorehead's critique, which appeared in the *American Historical Review* in fall 1913, praised the book as "evinc[ing] the most careful research and geologic skill. It does not seem to the writer that any person can controvert his observations. . . . we should have more and similar works along the same line elsewhere in the United States."[18]

Summer 1913, Winchell left his work in the lab at the old capitol and first traveled to Cambria to inspect Nickerson's dig. The dig had recently been the subject of a "moving picture film" by Pastime Amusement Company of Mankato, Minnesota. The movie showed Nickerson at the site, explaining important findings; "the photography is clear and sharp and every object stands out in strong relief." It showed "exactly what the eye witness would see watching the work."[19] The company thought the film had "considerable scientific value." For Winchell, who began his career with sailboats, horses and wagons, and birch-bark canoes, a motion picture of field research must have been both glamorous and questionable.

Then he and Upham took the train to the International Geological Congress in Toronto. Alfred Lacroix, traveling from Paris to attend the conference, thought Winchell looked "vigoreuz." The geologist-turned-archaeologist examined Paleolithic material in the university, then went on to Montreal and New York City, where he stopped off at Trenton, New Jersey, to see Dr. Abbott's work site. Winchell made a second trip to the East Coast four months later and delivered papers at both the Geological Society of America meeting in Princeton and the American Anthropological Association meeting in New York City. At the geological meeting, he met a wealth of old friends, men he had known for decades. Although John Stevenson would later observe that "at Princeton . . . he was clearly failing,"[20] most thought he looked hale and hearty. "I caught him going through the Geological Section of our Museum on his way to see Professor Putnam," noted John Woodworth of Harvard. "[He] looked so well preserved and vigorous that it never occurred to me that he was growing old."[21]

But something ailed the seventy-four-year-old. At Trenton, he asked Ernest Volk, a German-born archaeologist who was versed in herbal medicine, for treatment, and Volk recommended knotweed, *Polygonum aviculare.* Winchell had trouble procuring it once back in Minneapolis, so Volk sent him some of his own supply, gathered just the past summer, and also some *Equisetum,* common horsetail, which were used in combination for bladder complaints.[22]

The Christmas trip to the East Coast had filled Winchell with enthusiasm. His letters to Charles Abbott in Trenton bubbled over with ideas, research avenues to pursue, and thoughts on Abbott's findings. He had told the Historical Society that his work on Paleolithic people—even though his collecting would not be in Minnesota—would thrust the society into the forefront of national recognition, and that seemed to be coming true. And the specimens supporting his ideas were in Brower's collection, owned by the society.

Wishing to get a clearer idea of Winchell's plans, Upham, acting as society secretary, wanted to know his summer schedule. Winchell laid it out in six points: he wanted to visit northeastern Kansas to have another look at the chert terraces, and he hoped to have James Todd accompany him. On his way to Kansas, he intended to stop off in Cedar Falls, Iowa, to address the Iowa Academy of Sciences on "The Antiquity of Man." There were quarries in Missouri and Oklahoma to visit, and then he wanted to see collections at Springfield, Illinois. "In the work of studying and 'classifying' the Brower collection I have found some very old specimens of human fabrication found at East St. Louis (or near there) which came from New Jersey in pre-Wisconsin time." He wanted to know if they could also be found in Springfield's collection, for they would indicate pre-Wisconsin glacier trade between the East Coast and the Mississippi. And he wanted to go to Washington, D.C., in October to attend the International Congress of Americanists.[23]

Winchell's talk at Cedar Falls on April 24, 1914, went well. The Iowans warmed to his "genial, hearty manner . . . [and his] easy comradeship."[24] He had wanted to include Nickerson's work at Cambria in the talk, and Nickerson met him at the college, but he did not accompany him to the reception after the talk. Nickerson thought his boss was in good health, but there was apparently trouble that Winchell was not sharing, for instead of boarding the train to northeastern Kansas as planned, he headed home to Minneapolis.

ABOUT THIS TIME, LOTTIE WENT INTO Newton's study and took the slim, leather-bound volume they had called "Record of Illness" off the shelf. She caressed its smooth cover, thinking of the many times, and under such duress, that they had penned an entry in it. Newton had bought it in 1865—nearly fifty years ago—and the first entry had described her labor and delivery with Hortie. Newton had recorded the birth of every child in detail—except for Louise. Both of them had just been too busy to write after that birth. How long ago that was! How many times they had worried when the children were ill—with whooping cough, with mumps, with measles—thinking of how brother Alexander had lost four children to illness in Ann Arbor.

Caught within the leaves was a folded piece of paper. On it were her handwritten instructions, dictated by Dr. Stephenson in Adrian, of how to treat scarlet fever. They had come in handy once.

Lottie turned to the blank pages at the back of the book and headed her entry "Pa—Surgical Operation." She began to write.

She wrote of the prostatectomy performed by Charles Mayo in 1905. Nine years had passed since then, and she now described it in detail and observed that though the procedure "made him look and act perfectly well, he never really has been quite as strong nor has he been now for some time free from trouble causing frequent urination which is often bloody."[25]

She did not write, perhaps because it was not yet scheduled, that Winchell, with symptoms accompanying an enlarged prostate, would go into surgery again, on May 1 in Minneapolis at Northwestern Hospital. Although Charles Mayo thought he had removed the entire gland, perhaps he had not. Even today, the surgery is difficult because of the location of the gland.

This time, the surgeon was James Moore, a well-regarded Minneapolis physician. It is not known why Winchell did not return to Mayo. Perhaps the surgery seemed more straightforward this time around. Perhaps he felt there wasn't time.

He did not survive that surgery. He died the next day in a hospital room of stroke and kidney failure. It had been so quick. He saw the surgeon on Thursday, underwent surgery on Friday, and by Saturday evening, he was gone.

Newton Horace Winchell had been a solid and substantial rock for decades, and maybe because rocks seem so permanent, his friends and family were particularly stunned by his death. He was vital. He was vigorous. He was a force of nature, barreling through life. It would take them all a while to adopt the long view—geologists excel at the long view—and acknowledge this truth: that the fate of all rocks is dust.

EPILOGUE

As news of Newton Winchell's sudden death spread, there was an outpouring of grief and remembrance from geologists across the continent. Several adjectives kept appearing in these tributes: "genial," "vigorous," and "virile." At Winchell's funeral, former university president Folwell eulogized, recalling "the fine example of courtesy and manly dignity with which he adorned all his activities."[1] The presiding minister, J. Frank Stout, described him as "virile, aggressive, positive in his convictions, and strong where men needed strength."[2]

Inevitably, geologists, especially the older ones, recalled his brother Alexander, as well. Newton may have felt in the shadow of Alexander's charisma, particularly early in his career, but certainly over his long life, he blossomed into his own geologist, his own man.

"What clear, brilliant minds those two Winchells . . . had," remarked George H. Stone,[3] and Edward Dana of Yale, himself a geological name, echoed, "the Winchell name has been an honored one in the house for a great many years."[4] John Stevenson of New York University observed that "he did not make friends so readily as his brother Alexander did because a peculiar abruptness of delivery when on his feet led some to misunderstand him. . . . it was mainly a nervous manifestation, for he was a delightful man."[5]

It fell to Warren Upham to sum up his colleague and mentor's life, which he did in several ways, writing a biographical sketch for the Geological Society of America, a summary of his work in glacial geology and archaeology for *Economic Geology,* and a memorial to him for the Minnesota Academy of Sciences.[6] Upham concluded the GSA sketch with a simple acknowledgment: "He was my friend and it is hard to say Farewell!"[7]

Lottie survived her husband by twelve years, living until nearly ninety with daughter Louise in Minneapolis. In 1915, the University of Minnesota renovated the house at 120 State Street as housing for university women and named it the Charlotte Winchell cottage in recognition of her strong public support for women's higher education.[8]

Hortie, who with Ida Belle was out of the country at the time of his father's death, seeking treatment for his own failing health, would not live long. His career in mining took him far afield to China, Mexico, and Alaska; he and Ida Belle were in Petrograd, Russia, on business in 1917 at the start of the Bolshevik revolution. "Keep your eyes on Russia and your money out of it," he advised Minneapolis businessmen when he returned home.[9] He died in Los Angeles in 1923 of heart disease at age fifty-seven.

Ima was offered a job by the Dayton Company to lead the first salesmanship school in Minneapolis. It was a new venture in merchandising for Dayton's; she made many trips to New York City to learn from other schools. She later took a job organizing a similar school under the auspices of New York University and moved to the East Coast in 1919. Sadly, she died two weeks after her brother in 1923 at age fifty-six.

Avis and Uly Grant lived the remainder of their lives in Evanston, Illinois, in the two-story house that they built on the edge of Northwestern University. Uly was professor of geology at Northwestern from 1899 to his death in 1932, where he was beloved particularly for his summer field courses. Avis was active in civic and church affairs and died in 1964 at ninety-two.

Alexander Newton was professor of mineralogy and petrology at the University of Wisconsin–Madison from 1907 to 1944. His mineralogy course at the university had a profound effect on a youthful Sigurd Olson, who had been ready to drop what he saw as an essentially meaningless class. Alexander convinced him to stay, telling Olson that far from being "dead," minerals were "alive . . . each one a world and a universe in itself." He jarred Olson into a new appreciation for a living, dynamic planet.[10] Alex was best known for his textbooks on optical mineralogy, the first edition of which was written with his father, the last edition (fourth edition) with his son Horace, also a professor of geology, at Yale from 1945 to 1985. Alex died in 1958 at eighty-four.

Louise remained in Minneapolis. She and Draper settled into a comfortable home on Blaisdell Avenue. Sadly, Draper died at age forty-three in July 1923, a summer of grief for the Winchells, for Hortie and Ima would follow Draper within the month. Louise remarried in 1928 to Minneapolis businessman Burt Denman. She died in 1980 at age ninety-nine.

All of the family except for the Grants are buried in Lakewood Cemetery, Minneapolis, their graves clustered around a marker of polished gabbro.

Minnesota reinstated its geological survey in 1911, and the agency continues to this day as a research and service arm of the Newton Horace Winchell School of Earth Sciences at the University of Minnesota. They now occupy offices at 2609 Territorial Road in St. Paul. At present, the agency thrives with a major focus on comprehensive, three-dimensional mapping to fully understand the state's water resources, particularly groundwater.

The geology department at the University of Minnesota, eventually called "Newton Horace Winchell School of Earth Sciences," remained in Pillsbury Hall until 2017. In 2018, the legislature allocated funds to begin renovating the erstwhile science hall to house the English department, even though decades of geological material have yet to be removed.

The legacy of the natural history museum that Winchell played midwife to is embodied in the Bell Museum, recently moved to new quarters in St. Paul, near the University of Minnesota's St. Paul campus. The museum continues its mission first laid out in 1872, to educate the citizenry about the natural world.

The Minnesota Academy of Science continues in its mission to "promot[e] interest in and appreciation of science." The academy declined after Winchell's death in 1914 and disbanded in 1929. However, different people, seeing a need for an organization that advocated for science, reformed the academy in 1932. It was key to the establishment of the world-renowned Cedar Creek Ecosystem Science Reserve when it bought the initial five hundred acres of oak savannah in 1940. It hosts the Winchell Undergraduate Research Symposium each year concurrently with its annual meeting.

Many Minneapolis residents are familiar with the Winchell Trail, a two-and-a-half-mile walking path that hugs the west bank of the Mississippi River. It memorializes Winchell's great contribution to geology of accurately calculating the passage of time since the last glacier's retreat. A granite boulder with a commemorative plaque stands at the west end of the Franklin Avenue bridge.

Paddlers in the Boundary Waters Canoe Area Wilderness can canoe Winchell Lake, a long, narrow lake south of the Gunflint Trail in the eastern end of the BWCAW, named for Winchell by his brother Alexander. Many have accessed the BWCAW out of Fall Lake and into Newton Lake, also named by Alexander. Just west of Winchell Lake, Cherokee Lake once bore the name Ida Belle, named by her father. Farther west still, Ima Lake is nestled between gabbro ridges, accessible via Snowbank Lake. Outside of the Boundary Waters, in Lake Vermilion, Avis Island (now Hinsdale) is a large two-mile-long isle northwest of Frazer Bay.[11]

Both *Geology of Minnesota* and *The Aborigines of Minnesota* are considered foundational work by today's researchers. Time and scholarship have superseded them as definitive works of both fields.

George M. Schwartz, former director of the Minnesota Geological Survey, wrote: "Any attempt to do justice to the work of Newton Horace Winchell within a reasonable space is frustrated by the immensity of his accomplishment."[12] But let's try.

In additional to Winchell's calculation of the rate of ascent of St. Anthony Falls on the Mississippi River, he was the first to recognize, in the first year of the geological survey, that the Red River valley in northwestern Minnesota was the bed of a large glacial lake.[13] Later, as he examined the geology of northeastern Minnesota, he discovered that there were two formations rich in iron ore—the Vermilion, which was already known, and the Mesabi, the discovery of which foretold an immense capacity for mining in the state.[14] Furthermore, he considered the iron-bearing rock of the Mesabi to be part of the Taconic strata (now antiquated nomenclature) and called the rock "taconite."[15] It is fitting that Winchell Mountain from which he hailed is in the Taconic Range of New York State.

His claims of authenticity of the Kensington rune stone have not been borne out. Nor has his assessment of the stone artifacts of Kansas. But his insistence that geology can provide insight into archaeological questions now seems fresh and original. Geoarchaeology is a new field; its initial international conference was held in Saint John, New Brunswick, in 2005, although the geoarchaeology section of the GSA began earlier, in 1977. At the New Brunswick conference, Winchell's archaeological work was recognized: "In North America during . . . the early part of the twentieth century various earth scientists focused on determining the antiquity of humans in the New World—the possibility of an American Paleolithic. Leading figures in North American geology found the geological aspects of the early-man controversy compelling and made significant contributions to the resolution of questions about the glacial context of human remains. Prominent among them is the first state geologist of Minnesota, N. H. Winchell."[16]

ACKNOWLEDGMENTS

DESPITE MY DESIRE TO HAVE A GEOLOGICAL EYE, I was acutely aware of my shortcomings. When I began this project, I could say with a young Newton Winchell, “I know nothing about rocks!” (As a zoologist, I did know a bit more about fossils.) To prevent me from doing damage to myself and the discipline, my editor, Erik Anderson, coaxed Dr. Richard Ojakangas, retired geologist from the University of Minnesota Duluth (and University of Minnesota Press author), to serve as consultant. Dick was a faithful, conscientious reader of my drafts. If geological mistakes remain in this book, they are mine, not his.

Early in the project, I met Dr. David Southwick, former director of the Minnesota Geological Survey, over coffee and took notes; a month later, he followed up our conversation with a lengthy, detailed email. He shared what he knew about Winchell, laid down a basic history of geology, and gave me an inkling of how a geologist thinks. Many, many times in this project I went back to those two communications to find if I was on the right track, and I was amazed each time that Dave had already anticipated my questions.

Dr. Harvey Thorleifson, present director of the Minnesota Geological Survey, has been an enthusiastic supporter of the project. He was just an email message away whenever I had a question or needed an article that I couldn’t access online. When I’d ask if the survey had a particular journal in its library, he usually attached the article and sent it to me—within twenty-four hours.

Dr. Sally Kohlstedt, history of science professor embedded in the Department of Earth Sciences at the University of Minnesota, shared her perceptions and work on Winchell, especially on the geology museum. Her diligence unearthed two very good photographs of the museum from Winchell’s time, for which I am in her debt.

Dr. Michelle Terrell of Two Pine Resource Group is a working archaeologist and served as consultant for the two archaeology chapters. I know less about archaeology than I do about geology; she gave me a short course and corrected my misunderstandings. If errors remain, they are entirely mine.

Aaron Isaacs, my go-to for transportation information of the Twin Cities, supplied me, overnight, with the route and fare information for the trolley that trundled between Minneapolis and St. Paul in the early 1900s.

Newton and Lottie Winchell have numerous descendants scattered about the country. Louise was the only Winchell offspring to raise her children in Minneapolis and remain here. I met with Margie Dayton Ankeny, a granddaughter of Louise, and Jane Dayton Hall, Louise's daughter-in-law, both of Wayzata. My conversations with them were delightful, especially because these women are the living, breathing link to the illustrious Winchell clan. It was particularly interesting to learn that Margie, who sang professionally, possibly inherited the Winchell musical voice; others in the family are musical as well.

My editor at the University of Minnesota Press, Erik Anderson, provided insight and perspective at exactly the right times and encouragement when I was sure real geologists would hoot me out of the room. Associate editor Kristian Tvedten ferreted out amazing photographs and diligently paced the project. He embodies the idea that producing a book is a team effort. Copy editor Mary Keirstead, who produced flawless issues of two previous books of mine, once again kept me on the straight and narrow path. The entire production team at the University of Minnesota Press releases attractive, artistic books, and they continue to amaze me. Thank you to one and all.

My faithful writing group let me know that geological literacy is not widespread in educated society. Thanks to them this book is readable and aims for the curious reader who is interested in science and history. They read every single chapter and let me know when there was too much arcane detail and when I needed to enliven my subject. Past and present members are Kate Havelin, Judy Helgen, Patti Isaacs, Welcome Jerde, and Susan Narayan.

Finally, let's hear it for the guy with geological eyes, my husband, Tom Leaf, who read and commented on my chapters and once again was my medical consultant, this time on prostatectomies.

Thanks to every one of you! You guys rock! You've been so gneiss to me! I never took any of you for granite.

MEMBERS OF THE MINNESOTA GEOLOGICAL SURVEY

NEWTON WINCHELL HAD A DISCERNING EYE for talent, and many of the men he hired for the Minnesota Geological Survey went on to illustrious careers. They are listed here in alphabetical order.

Bayley, W. S. (1861–1943), worked on the survey 1890. He examined the Akeley Lake area, Hubbard County. He is known for his work on Minnesota's Pigeon Point and on Michigan's Menominee Iron Range for the U.S. Geological Survey (USGS).

Berkey, Charles P. (1867–1955), worked on the survey 1893–94. He was a geology professor at the University of Minnesota at the time he worked on the survey, then later went to Columbia University, where he was Newberry Professor of Geology. Considered a pioneer in engineering geology, he consulted on the construction of the Grand Coulee and Boulder Dams.

Dawson, George M. (1849–1901), worked on the survey 1885. Son of famed geologist John Dawson of McGill University, he led Geological Survey of Canada (GSC) parties to Canada's north and west and later headed the GSC from 1895 until 1901. Dawson City, Yukon, and Dawson Creek, British Columbia, are named after him.

Fairbanks, Harold W. (1860–1952), worked on the survey in 1887. He became an expert on gold belt geology in California and was later an advocate of geography education and an early voice for conservation.

Grant, Ulysses S. (1867–1932), worked on the survey 1883–88. He joined the geology department at Northwestern University in 1899 and became a beloved teacher. He served on the Geological and Natural History Survey of Wisconsin, examining copper deposits in northern Wisconsin; the Illinois Geological Survey; and the USGS.

Harrington, Mark W. (1848–1926), joined the survey in 1875. A graduate of the University of Michigan, he served as director of the Detroit

Observatory in Ann Arbor and, in 1891, became the first director of the U.S. Weather Bureau.

Herrick, Clarence L. (1858–1904), worked for the survey 1876–85. Among Minnesota zoologists, he was known for his monograph on freshwater invertebrates. He also reported on mammals, provided illustrations, and did an abstract of mining laws for the survey. After receiving his B.A. from the University of Minnesota, he took his doctorate at Leipzig, Germany, in neurobiology, then accepted a teaching post at Denison University in Ohio and later became president of the University of New Mexico.

Lawson, Andrew C. (1861–1952), worked on the survey 1891. A Canadian, he later joined the GSC, then became professor of mineralogy and geology at the University of California, Berkeley. He was a consulting geologist for the Golden Gate Bridge.

Lesquereux, Leo (1806–1889), consultant to the survey, 1883 and 1886. Swiss-born, he was an expert on mosses encountered studying European peat bogs. He was close friends with Louis Agassiz. After immigrating to the United States, he specialized in Carboniferous plants of Pennsylvania.

Oestlund, Oscar W. (1857–1948), worked for the survey 1885. He was an entomologist and lab assistant for the survey. An aphid expert, he trained at the University of Minnesota's University Farm (today, the St. Paul campus) and later taught at the university before becoming state entomologist. He was mentor to Edith Patch, the first professional woman entomologist, who credited him with proclaiming—sixty years before *Silent Spring*—that insects are beneficial in the large economy of nature.

Schuchert, Charles (1858–1942), worked on volume 3, part 1, of the survey's final report, identifying species of sponges, graptolites, corals, and brachiopods. He lived for a time at 120 State Street in 1891–93. He later joined the Yale University faculty and was director of Yale's Peabody Museum of Natural History from 1904 to 1923. His collection of brachiopods, now housed at the Peabody, was one of the world's largest.

Sperry, Lyman B. (1841–1923), worked on the survey summer 1877. He was a professor at Northfield College (now Carleton College) when he worked on the survey. He later spent thirteen years examining the area of Glacier National Park, where a glacier is named for him.

Spurr, Josiah E. (1870–1950), worked for the survey 1893, on the Mesabi Range rocks. He joined the USGS to map Alaska, where active volcano Mount Spurr is named for him. Later he traveled the world as a mining consultant.

Todd, James E. (1846–1922), worked on the survey 1892–93, examining Norman, Polk, Hubbard, Marshall, Roseau, Kittson, Beltrami, and part of Cass counties. Later he was state geologist for South Dakota and professor of geology at the University of Kansas, 1907–22. He received a personal letter from Charles Darwin nine days before Darwin's death in 1882.

Ulrich, Edward O. (1857–1944), came to the survey as a paleontologist. He described and illustrated bryozoans for the survey's eighth annual report. After many years of freelance paleontology work, he joined the USGS and continued there until retirement in 1932.

Upham, Warren (1850–1934), worked for the survey 1879–85. As a geologist, he is best known for his work on Glacial Lake Agassiz. Glacial Lake Upham in central Minnesota commemorates him. In a second career at the Minnesota Historical Society, he produced a comprehensive 718-page compendium, *Minnesota Place Names.*

Wadsworth, Marshman E. (1847–1921), worked on the survey 1886–87. He was the first president of Michigan Technological University at Houghton, Michigan, 1887–98.

Winchell, Alexander (1824–1891), worked on the survey field seasons 1886 and 1887 in northern Minnesota. He was the oldest brother of Newton Winchell. He was professor of geology at the University of Michigan, 1853–73 and 1879–91; as an educator, he was interested in popularizing geology and making its tenets available to the public.

Winchell, Alexander Newton (1874–1958). He was Newton Winchell's younger son and a professor of geology at the University of Wisconsin.

Winchell, Horace (1865–1923). Newton Winchell's older son accompanied the survey as a teenager and worked as a field geologist from 1885 to 1894. He later worked as a mining geologist for Anaconda Copper company and traveled the world as a freelance consultant.

NOTES

For countless hours, I read Winchell's personal journals and field journals. Whenever I felt I was losing the man inside all those words, I would write out a short scene in which I would make him walk and talk: I'd resurrect him. These proved oddly compelling, and I retained them in the book. Most are from Winchell's journals, which are housed at the Minnesota Historical Society. Several are from the reminiscences of his daughter Louise, and a boarder, Henry Beaudoux, found at the Hennepin County Library in Winchell's biography file. Details of weather and current events are from the *Minneapolis Tribune.*

1. FINDING A FOOTING

1. Newton Horace Winchell, journal, 3 August 1855, box 12, vol. 9, Newton Horace Winchell and Family Papers, Minnesota Historical Society (MHS), St. Paul.
2. Newton identified his book as "Fowler's Phrenology." The Fowler brothers, Lorenzo and Orson, based in New York City, were leading proponents of the pseudoscience and published profusely.
3. Warren Upham, "Memoir of Newton Horace Winchell," *Bulletin of the Geological Society of America* 26 (1915): 27.
4. Winchell, journal, 9 August 1855.
5. North East, New York, is categorized as a "town" in that state's organization but would be considered a "township" elsewhere. Spencer's Corner and Millerton are communities within North East.
6. Newton H. Winchell and Alexander N. Winchell, *The Winchell Genealogy: The Ancestry and Children of Those Born to the Winchell Name in America since 1635, with a Discussion of the Origin and History of the Name and the Family in England, and Notes on the Wincoll Family* (privately published by Horace V. Winchell, 1917), 108.
7. Winchell, journal, 17 August 1855.
8. Winchell, journal, 10 August 1855.
9. Winchell and Winchell, *The Winchell Genealogy,* 242–43.
10. Winchell, journal, 16 December 1855.
11. Winchell, journal, 30 September 1857.
12. Winchell, journal, 10 April 1856 and 25 April 1856.
13. Ibid.
14. Horace Winchell, "A Last Warning," completed June A.D. 1857, box 12, vol. 4, Newton Horace Winchell and Family Papers.
15. Winchell and Winchell, *The Winchell Genealogy,* 242–43.
16. Ibid.
17. Winchell, journal, 19 January 1857.

18. Newton Horace Winchell to Martin Winchell, 21 February 1857, box 1, Newton Horace Winchell and Family Papers.
19. Ibid.
20. Winchell, journal, 9 November 1855.
21. Winchell, journal, 27 October 1855.
22. Ibid.
23. Newton Horace Winchell to Martin Winchell, 21 February 1857.
24. Upham, "Memoir of Newton Horace Winchell," 27.
25. Winchell, journal, 13 September 1855.
26. Ibid.
27. Winchell, journal, 4 October 1855.
28. Winchell, journal, 3 November 1855.
29. Winchell, journal, 9 November 1855.
30. Winchell, journal, 12 November 1855.
31. Information about nineteenth-century district schools in New York State comes from Helen G. Trager, *The School House at Pine Tree Corner, North Salem, New York, 1784–1916: Teaching and Administrative Practices in a One-Room Rural School in Westchester County* (Harrison, N.Y.: Harbor Hill Books, for the North Salem Historical Society, 1976); some children in Dutchess County, New York, where the Winchell Mountain school was, attended the Pine Tree Corner school.
32. Ibid., 56–81.
33. Winchell, journal, 6 December 1855.
34. Ibid.
35. Trager, *The School House at Pine Tree Corner,* 45–46.
36. Winchell, journal, 12 March 1856.
37. Winchell, journal, 22 March 1856. Antoinette had just turned fifteen.
38. Winchell, journal, 10 April 1856.
39. Winchell, journal, 1 May 1856.
40. Winchell, journal, 12 October 1856.
41. Trager, *The School at Pine Tree Corner,* 82–106.
42. Winchell, journal, 12 December, 1856.
43. Trager, *The School at Pine Tree Corner,* 83.
44. Winchell, journal, 12 December, 1856.
45. Winchell, journal, 22 July 1856.
46. Winchell, journal, 19 January, 1857.
47. Newton Horace Winchell to Martin Winchell, 21 February 1857.
48. Winchell, journal, 20 March 1857.
49. Ibid.
50. Winchell, journal, 27 April 1857.

2. A SOLID FOUNDATION

1. Newton Horace Winchell, journal, 27 April 1857, box 12, vol. 9, Newton Horace Winchell and Family Papers, Minnesota Historical Society (MHS), St. Paul.
2. Chronological Development of the University of Michigan Campus, 1837–1970, "1850," http://umhistory.dc.umich.edu/mort/original/1850/1850.html.

3. Winchell, journal, 27 April 1857.
4. History of the University of Michigan, Department of Geology, before 1906, http://umhistory.s3-website.us-east-2.amazonaws.com/History_of_Geology.html.
5. Ibid.
6. "Biography," Alexander Winchell Papers, Bentley Historical Library, University of Michigan, Ann Arbor, quod.lib.umich.edu/b/bhlead/umich-bhl-86321?rgn=main;view=text. Eighty percent of oil deposits occur in anticlinal reservoirs.
7. Winchell, journal, 22 May 1857 and 6 June 1857.
8. Winchell, journal, 16 May 1857.
9. Winchell, journal, 30 September 1857, box 12, vol. 10.
10. Winchell, journal, 3 November 1857.
11. Ibid.
12. Winchell, journal, 14 August 1857.
13. Winchell, journal, 10 April 1858.
14. Winchell, journal, 1 October 1858.
15. Xenophon (430–354 B.C.E.) was a Greek historian. *Anabasis* was a military memoir.
16. Winchell, journal, 24 October 1858.
17. Winchell, journal, 4 December 1858.
18. Ibid.
19. Winchell, journal, 27 June 1859.
20. Winchell, journal, 10 July 1859.
21. Ibid.
22. Winchell, journal, 1 September 1859.
23. Winchell, journal, 23 March 1860.
24. Ibid.
25. Ibid.
26. Ibid.
27. Winchell, journal, 18 July 1861.
28. Ibid.
29. Fowler is better known as a prominent nineteenth-century phrenologist. His singular foray into architecture touted the energy-saving feature of the octagon, an approximation of a sphere, and anticipated the geodesic dome by a hundred years.
30. This was the second incarnation of the Michigan Geological Survey. The first was established in 1837, making it one of the oldest state surveys in the country. Douglass Houghton had served as its first director.
31. Wikipedia, s.v. "Asa Gray," Wikipedia.org/wiki/Asa_Gray. The publisher had requested a work that could be understood by lay people.
32. Winchell, journal, 8 May 1860.
33. Winchell, journal, 5 June 1860.
34. Ibid.
35. Winchell, journal, 18 May 1860.
36. Winchell, journal, 30 July 1861.
37. Winchell, journal, 5 June 1860.
38. Ibid.

39. Winchell, journal, June 1860, box 12, vols. 11, 12, and 13.
40. Winchell, journal, 29 June 1860, box 12, vol. 11. Some Mackinaw boats were equipped with retractable centerboards that gave them purchase in a wind.
41. Winchell, journal, 15 June 1860.
42. Ibid.
43. The account is taken from Winchell's journal, 3 July 1860, box 12, vol. 11.
44. Winchell, journal, 5 July 1860.
45. Winchell, journal, 15 July 1860.
46. Winchell, journal, 21 July 1860.
47. Ibid.
48. Winchell, journal, 24 July 1860.
49. Winchell, journal, 29 July, 1860.
50. Winchell, journal, 2 August 1860.
51. Winchell, journal, 20 August 1860. Newton was correct in his supposition.
52. Winchell, journal, 23 August 1860.
53. Winchell, journal, 2 January 1861.
54. Winchell, journal, 30 July 1861.
55. Ibid.
56. Ibid.
57. Winchell, journal, 6 December 1861.
58. Winchell, journal, 30 July, 1862. Winchell writes a graphic description of young Julius Alexander Winchell's death from diphtheria in his journal.
59. Winchell, journal, 6 December 1861.
60. Ibid.
61. Winchell, journal, 1 August 1862.
62. Ibid.
63. Winchell, journal, 28 June 1863.
64. Winchell, journal, 15 May 1869, box 12, vol. 10.
65. Ibid.
66. Winchell, journal, 15 April 1867, box 2, folder 1867.

3. METAMORPHOSIS

1. Newton Horace Winchell to Charlotte Sophia Imus, 8 July 1864, box 2, correspondence 1864–79, Newton Horace Winchell and Family Papers, Minnesota Historical Society (MHS), St. Paul.
2. Winchell to Imus, 12 July 1864.
3. Winchell to Imus, 26 July 1864.
4. Newton Winchell to Charlotte Winchell, 24 August 1871.
5. Winchell to Imus, 29 July 1864.
6. Ibid.
7. Newton Horace Winchell, journal, 24 August 1871, box 2, folder 1870–71, correspondence, Newton Winchell and Family Papers.
8. Winchell, journal, 31 [*sic*] September 1864, box 13, vol. 13, Newton Horace Winchell and Family Papers.

9. Winchell, journal, 21 January 1865.
10. James Dana (1813–1895) was a geologist; in 1865, he held the Silliman Professor of Natural History and Geology chair at Yale College.
11. Box 13, vol. 16, Newton Horace Winchell and Family Papers.
12. Winchell, journal, 28 June 1865.
13. Newton Winchell to Charlotte Winchell, 22 November 1865, box 2, correspondence 1864–1879.
14. Winchell, journal, 17 February 1866.
15. Ibid.
16. Newton Winchell to Charlotte Winchell, 18 October 1865.
17. Newton Winchell to Charlotte Winchell, 12 October 1865.
18. Newton Horace Winchell, "A Record of the Symptoms, Sickness and Treatment of Our Family in Every Case of Illness Occurring after Marriage August 24, 1864," box 13, vol. 14, Newton Horace Winchell and Family Papers.
19. The alternative to homeopathy was termed *allopathy* and was seen by some as treating merely the disease symptoms and not the underlying disharmony created by the disease. Allopathy could be harsh—as in bloodletting—but evolved into modern medicine.
20. Winchell, "A Record of the Symptoms."
21. There is no transcript at the University of Michigan recording Winchell's completion of the degree, but this is the only time period in which he could have accomplished this.
22. Winchell, journal, 9 May 1867.
23. Ibid.
24. Newton Winchell to Charlotte Winchell, 1 May 1867.
25. Newton H. Winchell and Alexander N. Winchell, *The Winchell Genealogy: The Ancestry and Children of Those Born to the Winchell Name in America since 1635, with a Discussion of the Origin and History of the Name and the Family in England, and Notes on the Wincoll Family* (privately published by Horace V. Winchell, 1917), 380, https://archive.org/details/winchellgenealog00winc/page/380.
26. Winchell was correct. The Ladies Library Association was the forerunner of the Adrian Public Library system; Shirley Ehnis, Adrian District Library, email message to author, 14 May 2015.
27. Winchell, journal, 6 January 1870.
28. Ibid.
29. N. H. Winchell is listed in the University of Michigan's alumni records as being awarded a master of arts degree in 1869, but the school no longer has transcripts for either the master's or the bachelor's degrees awarded in 1866; Michelle Henderson, University of Michigan Registrar's Office, email message to author, 15 May 2015.
30. The M.E. church in Ann Arbor followed the common practice of renting specific pews to members to generate funds for the congregation.
31. Winchell, 5 September 1869, box 2, folder 1868–69, correspondence.

4. WORKING GEOLOGIST

1. Newton Horace Winchell, field journal, 19 June 1870, box 13, vol. 20, Newton Horace Winchell and Family Papers, Minnesota Historical Society, St. Paul (MHS). A clinometer compass, essential for fieldwork, measures the dip (inclination) and strike (line along which inclination of bed runs) of sedimentary rock beds.
2. Winchell, field journal, 22 May 1870.
3. Winchell, field journal, 20 May 1870.
4. Ibid.
5. James Hutton, *Theory of the Earth* (1788), as quoted in Keith S. Thomson, "Vestiges of James Hutton," *American Scientist*, 89, no. 3 (May-June 2001), www.americanscientist.org/issues/pub/vestiges-of-james-hutton.
6. Horace Winchell, biblical manuscript, "A Last Warning," 1857, box 12, vol. 4, Newton Horace Winchell and Family Papers.
7. Charles J. Ammon, The Age of Earth Controversy, http://eqseis.geosc.psu.edu/~cammon/HTML/Classes/PhysicalGeology/Notes/SciUniversality/P07.html.
8. See Simon Winchester, *The Map That Changed the World: William Smith and the Birth of Modern Geology* (New York: Harper Collins, 2001).
9. Winchell pulled quotes from Dana's textbook and recorded them in a "Common-Place Book," "designed to assist students, professional men, and general readers in treasuring up knowledge for future use"; box 13, vol. 16, Newton Horace Winchell and Family Papers.
10. N. H. Winchell, "The Glacial Features of Green Bay of Lake Michigan, with Some Observations on a Probable Former Outlet of Lake Superior," *American Journal of Science and Arts* 2, nos. 7–12 (1871): 15–19, ed. James D. Dana and B. Silliman.
11. Winchell, journal, 10 December 1872, box 13, vol. 13, Newton Horace Winchell and Family Papers.
12. Ibid.
13. Newberry held a position as geologist at Columbia College (now Columbia University School of Mines) and maintained a residence in Cleveland, Ohio.
14. William E. Lass, "Minnesota's Quest for Salt," *Minnesota History* 52, no. 4 (winter 1990): 130–43.
15. Newton Horace Winchell to Charlotte Imus Winchell, 7 June 1871, box 2, folder 1864–79, correspondence, Newton Horace Winchell and Family Papers.
16. Ibid.
17. Newton Winchell to Charlotte Winchell, 7 April 1872.
18. Newton Winchell to Charlotte Winchell, 11 April 1872, box 2, folder 1872–75.
19. Ibid.
20. Newton Winchell to Charlotte Winchell, 17 April 1872.
21. Ibid.
22. Winchell, journal, 18 April 1872, box 2, folder 1872–75.
23. Winchell, journal, 14 April 1872. Newberry might have fueled Winchell's desire to see the Colorado River; Newberry had done exploratory work on it in 1857–58, but John Wesley Powell had recently returned from his historic trip down the river, and Winchell undoubtedly had heard some of his tales.

24. J. S. Newberry to W. W. Folwell, 12 April 1872, box 2, folder 1872–75, Newton Horace Winchell and Family Papers.
25. J. H. Klippert to unknown receiver, box 2, folder 1872–75.
26. Newton Winchell to Charlotte Winchell, 29 June 1872, box 2, folder 1872–1875 correspondence.
27. Newton Winchell to Charlotte Winchell, 26 September 1871, box 2, folder 1870–1871 correspondence.
28. Newton Winchell to Charlotte Winchell, 18 June 1872, box 2, folder 1870–71, correspondence.
29. Newton Winchell to Charlotte Winchell, 13 July 1872, box 2, folder 1872–75 correspondence.

5. SETTLING IN MINNEAPOLIS

1. *Minneapolis Sunday Tribune,* 22 September 1872, 1.
2. *Minneapolis Morning Tribune,* 17 September 1872, 3.
3. Lucile M. Kane, *The Falls of St. Anthony: The Waterfall That Built Minneapolis* (St. Paul: Minnesota Historical Society Press, 1987), 69–79.
4. Ibid., 82.
5. Lori Sturdevant, *The Pillsburys of Minnesota* (Minneapolis: Nodin Press, 2011), 43. The Pillsbury family owned two small mills on the west side of the river. The brand Pillsbury's Best XXXX Flour was first used 1 March 1873.
6. Ibid., 14.
7. Wikipedia, s.v. "Macalester College," Wikipedia.org/wiki/Macalester_College.
8. The congregation would take the name Our Lady of Lourdes, and the building is the longest-occupied church in Minneapolis.
9. F. Garvin Davenport,. "Newton H. Winchell, Pioneer of Science," *Minnesota History* 32, no. 4 (December 1951): 214–25.
10. Ibid.
11. Minnesota, *General Laws,* 1872, 86–88.
12. Aaron Isaacs, email message to author, 24 February 2016.
13. N. H. Winchell, *The Geological and Natural History Survey of Minnesota First Annual Report for the Year 1872* (Minneapolis: Johnson, Smith and Harrison, 1884), 20.
14. Ibid., 87–88.
15. Winchell, *The Geological and Natural History Survey of Minnesota,* 40.
16. Ibid., "Chart of Geological Nomenclature," between 40 and 41.
17. The founders of the Minnesota Academy of Natural Sciences were A. E. Ames, physician; A. F. Elliot, physician; S. C. Gale, lawyer; E. W. B. Harvey, superintendent of Minneapolis East District Schools; A. E. Johnson, physician; W. H. Leonard, physician; C. E. Rogers, physician; Charles Simpson, physician; M. D. Stoneman, dentist; A. W. Williamson, math teacher; and N. H. Winchell, geologist. The Minnesota Academy of Science is still in existence and holds an undergraduate research symposium each year to commemorate N. H. Winchell.
18. Chute Park and the Ard Godfrey house mark the site in 2016.
19. Sturdevant, *The Pillsburys of Minnesota,* 24.

20. Sanford Niles, *History and Civil Government of Minnesota* (Chicago: Werner School Book Co., 1897), 89.
21. University of Minnesota, 1874–75 calendar, 43–45, University of Minnesota Archives. Specifically, Winchell used Marcy's Sciopticon, which had been invented by Lorenzo Marcy of Philadelphia in 1872, so the device was cutting-edge technology.
22. Blowpipe analysis uses air passed through a flame, superheating it, to assay samples. The procedure was developed in the 1700s.
23. William E. Lass, "Minnesota's Quest for Salt," *Minnesota History* 52, no. 4 (winter 1990): 130.
24. Ibid., 137.
25. N. H. Winchell and S. F. Peckham, *The Geological and Natural History Survey of Minnesota: The Second Annual Report for the Year 1873* (Minneapolis: Harrison and Smith, 1893), 126–27.
26. Ibid., 143.
27. Winchell and Peckham, *The Geological and Natural History Survey of Minnesota,* 178. Winchell labels the limestone as Silurian in origin, but it is now considered Ordovician rock.
28. Ibid., 176–82.
29. However, later geological work has shown that the bluish clay is present in a three-to-six-foot-thick bed now known as the Blue Earth Siltstone, in the lower part of the Ordovician Dolomite. Richard Ojakangas, personal communication with author, 6 June 2016.
30. Winchell and Peckham, *The Geological and Natural History Survey of Minnesota,* 129.
31. This was Sioux Quartzite, which Winchell does not name. In Pipestone, Minnesota, the Pipestone County Museum, County Courthouse, and Carnegie Library are all built of Sioux Quartzite, as is the Van Dusen house in the Stevens Square neighborhood of Minneapolis.

6. FRACTURE

1. Caroline Winchell to Alexander Winchell, 19 October 1873, box 2, folder 1872–75, Newton Horace Winchell and Family Papers, Minnesota Historical Society (MHS), St. Paul.
2. "Newton Horace Winchell, "A Record of the Symptoms, Sickness and Treatment of Our Family in Every Case of Illness Occurring after Marriage August 24, 1864," 20 April 1874 entry, box 13, vol. 14, Newton Horace Winchell and Family Papers.
3. Ibid.
4. "10 Remedies Homeopaths Use to Treat Flu-like Symptoms," Homeopathy Plus!, http://homeopathyplus.com/10-homeopathic-remedies-that-treat-flu-like-symptom/.
5. Winchell, "A Record of the Symptoms."
6. N. H. Winchell, *The Geological and Natural History Survey of Minnesota: The Third Annual Report for the Year 1874,* 2d ed. (Minneapolis: Harrison and Smith, 1894), 145.

7. There is disagreement on this. G. B. Morey of the Minnesota Geological Survey in a historical article on the expedition uses this figure, but his source, *Custer's Gold* by Donald Jackson, claims the federal government did not allocate money for a geologist, and that Winchell's salary was paid by the University of Minnesota. Winchell, himself, states that the "state of Minnesota also may claim the honor of sending the first geologist through the unexplored interior of the Black Hills" in his 1874 annual report.
8. Donald Jackson, *Custer's Gold: The United States Cavalry Expedition of 1874* (New Haven, Conn.: Yale University Press, 1966), 2.
9. G. B. Morey, "Newton Horace Winchell, the George Armstrong Custer Expedition of 1874, and the 'Discovery' of Gold in the Black Hills, Dakota Territory, U.S.A.," *Earth Sciences History* 18, no. 1 (1999): 78–90.
10. Tim Brady, "General Custer and the Geology Professor," in *Gopher Gold: Legendary Figures, Brilliant Blunders, and Amazing Feats at the University of Minnesota* (St. Paul: Minnesota Historical Society Press, 2007), 56.
11. Jackson, *Custer's Gold,* 8–9.
12. Ibid., 9.
13. Evan S. Connell, *Son of the Morning Star: Custer and the Little Bighorn* (San Francisco: North Point Press, 1984), 236.
14. Jackson, *Custer's Gold,* 59.
15. Ibid., 51.
16. Morey, "Newton Horace Winchell," 82.
17. Newton Horace Winchell, field journal, 2 July 1874, box 1, folder Field Notebooks 1874, Newton Horace Winchell Papers, University of Minnesota Archives, Minneapolis.
18. Ibid.
19. Jackson, *Custer's Gold,* 26.
20. Morey, "Newton Horace Winchell," 81.
21. Jackson, *Custer's Gold,* 47.
22. Winchell, field journal, 16–17 July 1874.
23. Winchell, field journal, 22–23 July 1874.
24. Ibid.
25. Connell, *Son of the Morning Star,* 102–3.
26. Ibid., 113.
27. Winchell, field journal , 25 July 1874.
28. Ibid.
29. Jackson, *Custer's Gold.*
30. Winchell, field journal, 1 Aug. 1874.
31. Morey, "Newton Horace Winchell," 87.
32. *Chicago Times,* 13 March 1875; clipping in box 2, folder 1872–75, Newton Horace Winchell and Family Papers, MHS.
33. Brady, "General Custer and the Geology Professor," 61.
34. Quoted in Herbert Krause and Gary D. Olson, *Prelude to Glory: A Newspaper Accounting of Custer's 1874 Expedition to the Black Hills* (Sioux Falls, S.Dak.: Brevet Press, 1974), 237. Dogberry is a character in Shakespeare's *Much Ado about Nothing,* a bumbling night watchman with an inflated view of his own importance.

35. Account book #1, box 13, vol. 32, Newton Horace Winchell and Family Papers.
36. Winchell, "A Record of the Symptoms," 21 July 1875 entry.
37. Newton Horace Winchell to Alexander Winchell, 2 June 1875, box 2, folder 1872–75, Newton Horace Winchell and Family Papers.
38. Ibid.
39. Caroline Winchell to Alexander Winchell, 19 November 1873, box 2, folder 1872–75, Newton Horace Winchell and Family Papers.
40. See, for example, the 19 November 1873 letter cited in the previous endnote.
41. Newton Horace Winchell to Alexander Winchell, 2 June 1875, box 2, folder 42, Alexander Winchell Papers, Bentley Historical Library, University of Michigan, Ann Arbor.
42. Ibid.
43. Caroline Winchell to Alexander Winchell, 31 May 1875, box 2, folder 1872–75 correspondence, Newton Horace Winchell and Family Papers.
44. Winchell, "A Record of the Symptoms," undated, paragraph beginning "early in July, 1875."

7. BEDROCK AND RIVERS

1. F. Garvin Davenport, "Newton H. Winchell, Pioneer of Science," *Minnesota History* 32, no. 4 (December 1951): 214–25.
2. The house at 120 State Street was located where present-day Walter Library sits on the university mall; the hotel was where Wuling Hall is today.
3. Lori Sturdevant, *The Pillsburys of Minnesota* (Minneapolis: Nodin Press, 2011), 14.
4. Jane Pejsa, *Gratia Countryman: Her Life, Her Loves, and Her Library* (Minneapolis: Nodin Press, 1995), 46.
5. Mark W. Harrington became the first chief of the Weather Bureau, serving from 1891 to 1895; "History of the National Weather Service," National Weather Service, www.weather.gov/timeline.
6. W. E. Leonard, "Memorial for Newton Horace Winchell," *Bulletin of the Minnesota Academy of Sciences* 5, no. 2 (1914): 95–96.
7. N. H. Winchell, *The Geological and Natural History Survey of Minnesota: The Fourth Annual Report for the Year 1875* (St. Paul: Pioneer-Press Co., 1876), 16.
8. A brachiopod is any of a phylum of shelled invertebrates resembling a clam.
9. Winchell, *The Geological and Natural History Survey of Minnesota, Fourth Annual Report,* 62.
10. Ibid., 23.
11. Ibid., 11.
12. Allen Whitman, "Section VII. Entomology," in *The Geological and Natural History Survey of Minnesota: The Fifth Annual Report for the Year 1876,* by N. H. Winchell (St. Paul: Pioneer Press Co, 1877), 90–130. Interestingly, Whitman did not suggest insecticide to eliminate the pests, even though the field of organic chemistry was rapidly expanding in Germany. For a readable account, see W. W. Folwell, *A History of Minnesota,* vol. 3 (St. Paul: Minnesota Historical Society, 1926), 93–111.

13. Wikipedia, s.v. "Rocky Mountain locust," en.wikipedia.org/wiki/Rocky_Mountain_locust.
14. Theodore L. Hopkins, "Extinction of the Rocky Mountain Locust," review of *Locust: The Devastating Rise and Mysterious Disappearance of the Insect That Shaped the American Frontier,* by Jeffrey Lockwood (New York: Basic Books, 2004), *BioScience* 55, no. 1 (January 2005): 80–81, https://doi.org/10.1641/0006-3568(2005)055[0080:EOTRML]2.0.CO;2.
15. Horace Winchell, handwritten list, box 2, folder 1876–79, Newton Horace Winchell and Family Papers, Minnesota Historical Society (MHS), St. Paul.
16. Newton Horace Winchell to Charlotte Imus Winchell, 26 June 1878, box 2, folder 1876–79, Newton Horace Winchell and Family Papers.
17. Newton Winchell to Charlotte Winchell, 19 July 1876, box 2, folder 1876–79.
18. Newton Winchell to Charlotte Winchell, 22 July 1876, box 2, folder 1876–79.
19. Some counties were created after 1876. Today Minnesota has eighty-seven counties.
20. Lucile M. Kane, *The Falls of St. Anthony: The Waterfall That Built Minneapolis* (St. Paul: Minnesota Historical Society Press, 1987).
21. See, for example, Winchell, *The Geological and Natural History Survey of Minnesota: Fourth Annual Report,* 28.
22. Richard W. Ojakangas, *Roadside Geology of Minnesota* (Missoula, Mont: Mountain Press Publishing, 2009), 251.
23. Winchell, *The Geological and Natural History Survey of Minnesota: Fifth Annual Report.*
24. Ibid., 188.
25. Richard W. Ojakangas and Charles L. Matsch, *Minnesota's Geology* (Minneapolis: University of Minnesota Press, 1982), 108.
26. F. W. Sardenson, "Glacier Geology Work of Professor N. H. Winchell," *Minnesota Academy of Natural Science Bulletin* 5 (1914): 89–92.
27. Ibid.
28. Ibid.; see also Warren Upham, "Memoir of Newton Horace Winchell," *Bulletin of the Geological Society of America* 26 (1915): 30; and George Schwartz, "Newton Horace Winchell: A Tribute," 2012, Newton Horace Winchell School of Earth and Environmental Sciences, University of Minnesota, www.esci.umn.edu/winchell.

8. THE WINCHELLS BROADEN THEIR REACH

1. "Ward's National Science Establishment," RocWiki: The People's Guide to Rochester, rocwkik.org/Ward's_Natural_Science_Establishment.
2. "Must Museum Be Closed Continually?," *Ariel* 1, no. 1 (1877): 2, University of Minnesota student newspaper, University of Minnesota Archives, Minneapolis.
3. N. H. Winchell, *The Geological and Natural History Survey: The Fifth Annual Report for the Year 1876* (St. Paul: Pioneer Press Company, 1877), 7.
4. N. H. Winchell, *The Geological and Natural History Survey of Minnesota: The Seventh Annual Report for the Year 1878* (Minneapolis: Johnson, Smith and

Harrison, 1879), 81–123, plus illustrations. Herrick later was professor of geology and natural history at Denison University, and the second president of the University of New Mexico.

5. Newton Winchell to Charlotte Winchell, 12 November 1876, box 2, folder 1876–79, Newton Horace Winchell and Family Papers, Minnesota Historical Society (MHS), St. Paul.
6. Newton Winchell to Charlotte Winchell, 3 November 1876.
7. Newton Winchell to Charlotte Winchell, 29 October 1876.
8. Ibid.
9. Barbara Stuhler, *Gentle Warriors: Clara Ueland and the Minnesota Struggle for Woman Suffrage* (St. Paul: Minnesota Historical Society Press, 1995), 27.
10. Charlotte Winchell, from an account of the historic election, undated, box 10, folder articles and speeches, Charlotte (Mrs. Newton H.), 1870–96, Newton Horace Winchell and Family Papers.
11. Newton Winchell to Charlotte Winchell, 29 October 1876, box 2, folder 1876–1879.
12. "The East Side School Board: How the Ladies Looked and Conducted Themselves in Their New Positions," *Evening Mail,* 10 April 1876, a clipping interspersed between the letters, box 2, folder 1876–79, Newton Horace Winchell and Family Papers.
13. Postcard dated 10 April 1876, box 2, folder 1876–79, Newton Horace Winchell and Family Papers.
14. *Ariel* 1, no. 2: 21, University of Minnesota Archives.
15. *Ariel* 1, no. 4 (1 February 1878): 42.
16. T. S. Roberts to Frank Benner, 3 February 1878, box 1, Thomas S. Roberts and Family Papers, MHS, St. Paul.
17. "Home Hits and Happenings," *Ariel* 1, no. 4 (1 February 1878): 43.
18. *Ariel* 1, no. 9 (6 June 1878): 90.
19. Ibid., 97–98.
20. N. H. Winchell, "Annual Address of the Retiring President," 1880, *Bulletins of the Minnesota Academy of Natural Sciences* 1 (1873–79): 389–401.
21. Ibid., 392.
22. Ibid.
23. Ibid., 400.
24. Ibid.
25. N. H. Winchell, "The State and Higher Education," 1881, *Bulletin of the Minnesota Academy of Natural Sciences* 2, no. 3 (April 1881): 45–62.
26. Ibid., 45.
27. Ibid., 46.
28. Ibid., 57.
29. Ibid., 61.
30. "Newton H. Winchell," 2, Minnesota Geological Survey, www.mngs.umn.edu/Newton.pdf.
31. Newton Winchell to Charlotte Winchell, 6 October 1877, box 2, folder 1876–1879.
32. As an old man, Juni wrote an engaging memoir of his experience, *Held in Captivity* (New Ulm: Liesch-Walter Printing Co., 1926).

33. Newton Winchell to Charlotte Winchell, 6 October 1877.
34. Ibid.

9. ROCKS OF FIRE

1. Richard W. Ojakangas, personal communication with author via email, 23 November 2016.
2. In Anishinaabemowin, Chief Moquabimetem.
3. Winchell does not identify this carrier by name. John Beargrease began shuttling mail along the shore in 1879; his father, Chief Beargrease, also ran the mail and had several carriers under him. Daniel Lancaster, *John Beargrease: Legend of Minnesota's North Shore* (Duluth, Minn.: Holy Cow! Press, 2009), 34.
4. Newton Horace Winchell to Charlotte Imus Winchell, 29 June 1878, box 2, folder 1876–79, Newton Horace Winchell and Family Papers, Minnesota Historical Society (MHS), St. Paul.
5. Newton Winchell to Charlotte Winchell, 10 November 1876.
6. Newton Winchell to Charlotte Winchell, 14 July 1878.
7. Newton Winchell to Charlotte Winchell, 31 July 1878.
8. Newton Horace Winchell, journal, 14 June 1858, box 12, vol. 10, Newton Horace Winchell and Family Papers.
9. Newton Winchell to Charlotte Winchell, 23 June 1878, box 2, folder 1876–79.
10. Newton Winchell to Charlotte Winchell, 14 July 1878.
11. Newton Winchell to Charlotte Winchell, 20 June, 26 June, and 17 July 1878.
12. Newton Winchell to Charlotte Winchell, 23 June 1878.
13. Newton Winchell to Charlotte Winchell, 26 June 1878.
14. Newton Winchell to Charlotte Winchell, 30 June 1878.
15. Newton Winchell to Charlotte Winchell, 14 July 1878.
16. Ibid.
17. N. H. Winchell, *The Geological and Natural History Survey of Minnesota: The Seventh Annual Report for the Year 1878* (Minneapolis: Johnson, Smith and Harrison, 1979), 9.
18. N. H. Winchell, comment after specimen 109A, field notes, 9 June–22 July 1878, box 14, vol. 37, Newton Horace Winchell and Family Papers. Woodland caribou, the native ungulate on the North Shore, would disappear before 1940, as logging caused vegetation that favored white-tailed deer over caribou.
19. Winchell, comment under specimen 112, field notes, 9 June–22 July 1878.
20. Richard W. Ojakangas writes, "Winchell's interpretation was largely correct. Today his 'feldspar' is the rock anorthosite which is comprised of anorthite feldspar, calcium-rich plagioclase. The coarse-grained anorthosite . . . crystallized at depth and was carried to the surface as inclusions in magma"; personal communication with author via email, 30 December 2016.
21. Lancaster, *John Beargrease,* 23–24.
22. Tony Dierckins, with Linda Conradi, "Beaver Bay," Zenith City Press, http://zenithcity.com/archive/north-south-shore/beaver-bay/.
23. Winchell, commentary near specimen 138, field notes, 9 June–22 July 1878.
24. Winchell, comment at specimen 140, field notes, 9 June–22 July 1878.

25. Winchell, comment after rock specimen 171, field notes, 31 July–22 August 1878, box 14, vol. 39.
26. N. H. Winchell, *The Minnesota Geological and Natural History Survey: The Tenth Annual Report for the Year 1881* (St. Paul: J. W. Cunningham, 1882), 60. A total solar eclipse was observed 29 July 1878 in the United States from Wyoming to Texas.
27. Winchell, *The Geological and Natural History Survey of Minnesota: Seventh Annual Report,* 28.
28. Ibid.; also T. S. Roberts, journal entry, 27 July 1879, in *Shotgun and Stethoscope: The Journals of Thomas Sadler Roberts,* ed. Penelope Krosch (Minneapolis: James Ford Bell Museum of Natural History University of Minnesota, 1991), 147–48. Roberts did not name Mrs. Howenstein in his account.
29. Roberts, journal entry, 27 July 1879, in *Shotgun and Stethoscope,* 148.
30. Ibid., 151.
31. Winchell, *Minnesota Geological and Natural History Survey: Seventh Annual Report,* 9.
32. In a letter dated 24 July 1879, Newton refers to Lottie having seen Lake Superior, so the vacation apparently came off.
33. Mark J. Severson, "The History of Gold Exploration in Minnesota," Technical Report of the Natural Resources Research Institute of the University of Minnesota-Duluth, October 2011, http://hdl.handle.net/11299/187169.

10. PORTAGES

1. N. H. Winchell, *The Geological and Natural History Survey of Minnesota: The Ninth Annual Report for the Year 1880* (St. Peter: J. K. Moore, 1881), 27.
2. Ibid., 28.
3. Ibid., 71.
4. Ibid.
5. Ibid., 72.
6. Newton Winchell to Charlotte Winchell, 16 September 1878, box 2, folder correspondence 1876–79, Newton Horace Winchell and Family Papers, Minnesota Historical Society (MHS), St. Paul.
7. Ibid.
8. N. H. Winchell, *The Geological and Natural History Survey of Minnesota: The Seventh Annual Report for the Year 1878* (Minneapolis: Johnson, Smith and Harrison, 1879), 18.
9. Ibid., 20.
10. Richard W. Ojakangas, *Roadside Geology of Minnesota* (Missoula, Mont.: Mountain Press Publishing, 2009), 173.
11. Winchell, *The Geological and Natural History Survey of Minnesota: Ninth Annual Report,* 85.
12. Ibid., 87.
13. Ibid., 88.
14. Warren Upham, *Minnesota Place Names: A Geographical Encyclopedia,* 3rd ed. (St. Paul: Minnesota Historical Society, 2001), 310.

15. Winchell, *The Geological and Natural History Survey of Minnesota: Ninth Annual Report,* 92.
16. Ibid., 98.
17. Ibid., 101–2.
18. Ibid., 108.
19. Winchell, *The Geological and Natural History Survey of Minnesota: Seventh Annual Report,* 23.
20. Winchell, *The Geological and Natural History Survey of Minnesota: Ninth Annual Report,* 109.
21. Ibid., 110–11.
22. Newton Winchell to Charlotte Winchell, 25 October 1878, box 2, folder correspondence 1875–79.
23. Nathan Butler, "Tributes [of Newton Horace Winchell] from Early Associates and Pupils," *Bulletin of the Minnesota Academy of Sciences* 5, no. 2 (July 1914): 93.

11. BACK TO THE NORTH SHORE

1. N. H. Winchell, *The Geological and Natural History Survey of Minnesota: The Eighth Annual Report for the Year 1879* (St. Paul: Pioneer Press Company, 1880), 10.
2. Ibid.
3. Ibid., 85.
4. Encyclopaedia Britannica, s.v. "Louis Agassiz," https://www.britannica.com/biography/Louis-Agassiz.
5. Winchell's field camps often hosted visitors who aided the survey. In 1879, state senator H. B. Wilson accompanied Winchell in May; Adin and Den Brooks, university students, were at the June camp at Fond du Lac, as was their father, Professor Jabez Brooks; later, Professor Geo. Weitbrecht, St. Paul Schools, accompanied Hall to Grand Marais. T. S. Roberts states specifically that he needed to pay his way (T. S. Roberts to Frank Benner, 27 June 1878, box 1, Thomas S. Roberts and Family Papers, Minnesota Historical Society [MHS], St. Paul). All other references: box 2, folder 1876–79, Newton Horace Winchell and Family Papers, MHS. Spending time in camp was a means to getting "fresh air," a recommended cure for tuberculosis; Terry and Adin Brooks were both ill in 1879.
6. Newton Winchell to Charlotte Winchell, 11 May 1879, box 2, folder 1876–79, Newton Horace Winchell and Family Papers.
7. Ibid.
8. Ibid.
9. This section of the St. Louis River is now protected by Jay Cooke State Park.
10. Newton Winchell to Charlotte Winchell, 23 June 1879.
11. N. H. Winchell, *The Geological and Natural History Survey of Minnesota: The Tenth Annual Report for the Year 1881* (St. Paul: J. W. Cunningham, 1882), 7.
12. Ibid., 6.
13. T. S. Roberts to John Roberts, 28 July 1879, box 1, folder "Correspondence 1877–1880," Thomas S. Roberts and Family Papers.

14. N. H. Winchell, *The Geological and Natural History Survey of Minnesota: The Eighth Annual Report for the Year 1879* (St. Paul: Pioneer Press Company, 1880), 24.
15. Ibid., 7.
16. Newton Winchell to Charlotte Winchell, 22 June 1879.
17. Winchell, *The Geological and Natural History Survey of Minnesota: Tenth Annual Report,* 49.
18. Northwest of Siskiwit Bay, the site is now a national park campground.
19. Winchell, *The Geological and Natural History Survey of Minnesota: Tenth Annual Report,* 51. The Island Mine Co., formed in 1873, employed 130 people that year, had disastrous fires in 1874, and closed in 1875.
20. Ibid., 54.
21. Newton Winchell to Charlotte Winchell, 17 August 1879. Archaeologists today know "the ancients" were Paleo-Indians who mined the copper about five thousand years ago.
22. Now the site of Sleeping Giant Provincial Park in Ontario, Canada.
23. James H. Marsh, "Silver Islet," last edited March 4, 2015, Canadian Encyclopedia, www.thecanadianencyclopedia.ca/en/article/silver-islet/. Richard W. Ojakangas, personal communication with author via email, 30 December 2016: "The mine bottomed at 1260 ft. The reason it is an islet rather than an island is that it was only 80 ft. across, and 2 ½ ft. above lake level. It was expanded by hauling in rock and building sea walls. . . . It is an Ontario Archeological and Historical Site. . . . The owner of the mine was A. H. Sibley, brother of Minnesota's first governor, Henry Hastings Sibley."
24. T. S. Roberts to John Roberts, 28 July 1879.
25. "Report of Professor C. W. Hall," in Winchell, *The Geological and Natural History Survey of Minnesota: Eighth Annual Report,* 131, 161.
26. Ibid., 155; Roberts's reports can be found on page 138 (plants) and page 155 (birds) of the *Eighth Annual Report.*
27. Winchell, *The Geological and Natural History Survey of Minnesota: Eighth Annual Report,* 133.
28. Ibid.
29. The state would not act in time, and today, fisheries biologists work to reestablish the native population in North Shore rivers; Chel Anderson and Adelheid Fischer, *North Shore: A Natural History of Minnesota's Superior Coast* (Minneapolis: University of Minnesota Press, 2015), 100–8.
30. Winchell, *The Geological and Natural History Survey of Minnesota: Eighth Annual Report,* 135.
31. Ibid., 134.
32. Ibid., 135.
33. Ibid.
34. Ibid.
35. Ibid., 129.
36. Ibid., 13.
37. Ibid.

12. THE BOOM

1. "Handbook of Minneapolis, Prepared for the Thirty-Second Annual Meeting of the American Association for the Advancement of Science," Minneapolis, 15–22 August 1883, 58.
2. T. S. Roberts, journal entry for 2 May 1878, in *Shotgun and Stethoscope: The Journals of Thomas Sadler Roberts,* ed. Penelope Krosch (Minneapolis: James Ford Bell Museum of Natural History University of Minnesota, 1991), 109.
3. Lori Sturdevant, *The Pillsburys of Minnesota* (Minneapolis: Nodin Press, 2011), 85.
4. Ibid., 91.
5. "Handbook of Minneapolis," 59.
6. Ibid., 62.
7. N. H. Winchell, *The Geological and Natural History Survey of Minnesota: The Ninth Annual Report for the Year 1880* (St. Peter: J. K. Moore, 1881), 314.
8. Ibid., 344.
9. Ibid., 187.
10. Ibid., 254, 257.
11. N. H. Winchell, *The Geological and Natural History Survey of Minnesota: The Seventh Annual Report for the Year 1878* (Minneapolis: Johnson, Smith and Harrison, 1879), 7.
12. Newton H. Winchell, "Building Stone and Scattered Misc. notes," box 14, vol. 42, Newton Horace Winchell and Family Papers, Minnesota Historical Society (MHS), St. Paul. Concrete was not widely used at the time.
13. *Ariel* 8, no. 1, 2 (October 1884): 7, https://catalog.hathitrust.org/Record/008375794.
14. "Home Hits and Happenings," *Ariel* 7, no. 9 (May 14, 1884): 159, https://catalog.hathitrust.org/Record/008375794.
15. Handbills of various musical productions, box 1, folder 1886–89, Gratia Countryman and Family Papers, MHS.
16. The organization is called a "fraternity" but is comprised of women.
17. Countryman would become a nationally known librarian and lead the Minneapolis Public Library for over thirty years.
18. Newton Horace Winchell, "A Record of the Symptoms, Sickness and Treatment of Our Family in Every Case of Illness Occurring after Marriage August 24, 1864," box 13, vol. 14, Newton Horace Winchell and Family Papers.
19. Warren Upham, *Minnesota Place Names: A Geographical Encyclopedia,* 3rd ed. (St. Paul: Minnesota Historical Society Press, 2001), 8.
20. T. B. Walker, N. H. Winchell, and A. F. Elliot to American Association for the Advancement of Science (AAAS), 4 January 1883, AAAS local committee records, box 14, vol. 43, Newton Horace Winchell and Family Papers.
21. David A. Lanegran and Ernest R. Sandeen, *The Lake District of Minneapolis: A History of the Calhoun-Isles Community* (Minneapolis: University of Minnesota Press, 2004), 16.
22. "Handbook of Minneapolis," 117.
23. *Minnesota Daily Tribune,* 22 August 1883, 2.

24. Ibid. Lesley, of Philadelphia, had been trained as a minister, and his expertise was in fossil fuels.
25. "Scientists at Taylors Falls," *Minnesota Daily Tribune,* 24 August 1883, 6.
26. N. H. Winchell, *The Geology of Minnesota,* vol. 1 (Minneapolis: Johnson, Smith and Harrison, 1884), 8, 5, 22.
27. Preface to "Handbook of Minneapolis," 1.
28. [Newton Horace Winchell?], "A History of the First Methodist Episcopal Church of Minneapolis," [1884?], box 1, Folder Correspondence and Miscellaneous Papers, 1884–89, First Methodist-Episcopal Church (Minneapolis, Minn.) Papers, MHS. Note: This document is not signed nor dated, but it is in Winchell's handwriting, and he was the church historian.

13. FIELDWORK, POLITICS, AND FEMINISM

1. Theodore C. Blegen, "Red Earth, Iron Men, and Taconite," in *Minnesota: A History of the State* (University of Minnesota Press, 1975), 364.
2. Ibid.
3. Alexander Winchell, "Report of Geological Observations Made in Northeastern Minnesota during the Season of 1886," in *The Geological and Natural History Survey of Minnesota: The Fifteenth Annual Report for the Year 1886,* by N. H. Winchell (St. Paul: Pioneer Press Company, 1887), 13.
4. Ibid., 14.
5. Ibid., 13.
6. Ibid., 15.
7. Ibid., 37.
8. Warren Upham, *Minnesota Place Names: A Geographical Encyclopedia,* 3rd ed. (St. Paul: Minnesota Historical Society, 2001), 315.
9. Winchell, *The Geological and Natural History Survey of Minnesota: Fifteenth Annual Report,* 98.
10. Ulysses S. Grant to Avis Winchell, 17 July 1887, box 1, folder 5, correspondence, Ulysses Sherman Grant to Avis Winchell, Ulysses Sherman Grant Papers, Northwestern University Archives, Evanston, Ill.
11. Uly Grant's full name was Ulysses Sherman Grant, not related to the president, Ulysses Simpson Grant.
12. Ulysses S. Grant to Avis Winchell, August 1, 1888, box 1, folder 6, correspondence, Ulysses Sherman Grant to Avis Winchell, Ulysses Sherman Grant Papers.
13. Ulysses S. Grant to Avis Winchell, 8 July 1888.
14. Ulysses S. Grant to Avis Winchell, 25 August 1888.
15. Ulysses S. Grant to Avis Winchell, 8 July 1888.
16. Ulysses S. Grant to Avis Winchell, 17 June 1888.
17. Wikipedia, s.v. "Women's Christian Temperance Union," www.en.wikipedia.org/wiki/Woman%27s_Christian_Temperance_Union.
18. C. S. Winchell, "Report of the State Superintendent of Scientific Temperance Instruction," in *A Brief History of the Minnesota Women's Christian Temperance Union from Its Organization, September 6, 1877 to 1939,* by Bessie Scovell (St. Paul: Bruce Publishing, 1939), 187.
19. Ibid., 187–92.

20. Ibid., 190.
21. Ulysses S. Grant to Avis Winchell, 8 July 1888, box 1, folder 6, correspondence 1888, Ulysses Sherman Grant to Avis Winchell, Ulysses Sherman Grant Papers.
22. Mary L. Blbnchard [*sic*], "Prophecy for '88," *Ariel* 11, no. 9 (15 May 1888): 128.
23. Ima C. Winchell, "Woman," *Ariel* 11, no. 9 (15 May 1888): 132.
24. Unattributed, but probably Louise Winchell Dayton Denman, "Life with Father," undated, envelope 2, Newton Horace Winchell and Alexander N. Winchell Papers, Biography Files, Hennepin County Library Special Collections. Note that the Geological Society of Minnesota incorrectly attributes this to Avis Winchell Grant.
25. "Story of a life—Mrs. Burt J. Denman," undated, unsourced, from the file of Jane Hall, interview by author, 16 January 2017, Wayzata, Minn.
26. T. A. Rickard, "Horace V. Winchell, Mining Geologist," interview, reprint from *Mining and Scientific Press,* 1919, https://catalog.hathitrust.org/Record/003203468.

14. SHAPING A SCIENCE

1. [N. H. Winchell?], "The Prenatal History of the Geological Society of America," *American Geologist* 6, no. 3 (September 1890): 181–92.
2. Ibid., 191.
3. For example, see Newton H. Winchell to Alexander Winchell, 21 February 1888, box 2, folder 63, correspondence January–March 1888, Alexander Winchell Papers, Bentley Historical Library, University of Michigan, Ann Arbor.
4. The seven original editors were Samuel Calvin, University of Iowa, Iowa City; Edward Claypole, Bechtel College, Akron, Ohio; Persifor Frazer, the Franklin Institute, Philadelphia; Lewis Hicks, University of Nebraska, Lincoln; Edward Ulrich, Illinois Geological Survey; Alexander Winchell, University of Michigan, Ann Arbor; and Newton Winchell, University of Minnesota, Minneapolis.
5. "Home Hits," *Ariel* vol. 11, no. 3 (December 1, 1887): 33.
6. N. H. Winchell, "A Proposed Geological Society," *American Geologist* 1, no. 6 (June 1888): 394.
7. N. H. Winchell, *The Geological and Natural History Survey of Minnesota: The Eighth Annual Report for the Year 1879* (St. Paul: Pioneer Press Company, 1980), 173–75.
8. Ibid.
9. Donald Worster, *A River Running West: The Life of John Wesley Powell* (New York: Oxford University Press, 2001), 160, 166.
10. Ibid., 199–200.
11. Ibid., 115, 114.
12. Ibid., 117, 126.
13. Newton H. Winchell to Alexander Winchell, 21 February 1888, box 2, folder 63 correspondence January–March 1888, Alexander Winchell Papers.
14. Ibid.
15. Newton H. Winchell to J. W. Powell, 30 April 1883, box 3, folder correspondence

1880–84, Newton Horace Winchell and Family Papers, Minnesota Historical Society (MHS), St. Paul.
16. Newton H. Winchell to Alexander Winchell, 21 February 1888.
17. Ibid.
18. Ibid.
19. The GSA names these original founders: H. L. Fairchild, Rochester University, N.Y.; James Hall, State Museum, Albany, N.Y.; Charles H. Hitchcock, Dartmouth College; J. F. Kemp, Cornell University; W. J. McGee, USGS; H. B. Nason, Rensselaer Polytechnic Institute, Troy, N.Y.; J. J. Stevenson, University of the City of New York; I. C. White, West Virginia University, Morgantown; H. S. Williams, Cornell University, Ithaca, N.Y.; J. F. Williams, Pratt Technical Institute, Brooklyn, N.Y.; S. G. Williams, Cornell University, Ithaca, N.Y.; Alexander Winchell, University of Michigan, Ann Arbor; and N. H. Winchell, University of Minnesota, Minneapolis.
20. Justin M. Samuel, "The Geological Society of America and Its Founders – John Wesley Powell," blog, Geological Society of America, June 26, 2013, https://community.geosociety.org/blogs/justin-samuel/2013/06/26/the-geological-society-of-america-and-its-founders-john-wesley-powell.
21. In February 1889, *The American Geologist* (3, no. 2) fixes the name as "Geological Society of America" at the December 1888 meeting, but due to constitutional technicalities, the name did not become official until December 1889. See Arthur Mirsky, *The Founding of the Geological Society of America: A Retrospective on Its Centennial Birthday, 1888–1988* (Boulder, Colo.: History of Geology Division, Geological Society of America, 1988), 6.
22. Ibid., 31.
23. Ibid., 41.
24. Edwin Butt Eckel, *The Geological Society of America: Life History of a Learned Society,* Memoir 155 (Boulder, Colo.: Geological Society of America, 1982), 11.
25. "Editorial Comment," *American Geologist* 3, no. 2 (February 1889): 140–46.
26. "Personal and Scientific News," *American Geologist* 5, no. 2 (February 1890): 122.
27. Larry Millett, *Lost Twin Cities.* (St. Paul: Minnesota Historical Society Press, 1992), 67. Despite the building's elegance, Millett terms it a "lemon."
28. Ibid., 115.
29. Ibid., 49.
30. "Pillsbury Hall," Department of Earth and Environmental Sciences, https://www.esci.umn.edu/facilities/pillsbury.
31. "Home Hits and Happenings," *Ariel* 12, no. 9 (May 21, 1889): 185.
32. N. H. Winchell, *The Minnesota Geological and Natural History Survey: The Eighteenth Annual Report for the Year 1889* (Minneapolis: Minnesota Geological Survey, 1890), 4.
33. Ibid.
34. Winchell, *The Minnesota Geological and Natural History Survey: The Eighth Annual Report,* 27–30.
35. Ibid. A specimen on loan from the University of Minnesota remains on display at the Estherville Public Library in 2017.
36. "A Catalogue of the Meteorites in the University Collection," in *The Minnesota*

Geological and Natural Survey: The Nineteenth Annual Report for the Year 1890, by N. H. Winchell (Minneapolis: Harrison and Smith, 1892), 170–92.

37. N. H. Winchell, "Natural Science at the University of Minnesota," *American Geologist* 3, no. 3 (March 1889): 165–69.
38. "Pillsbury Hall," Department of Earth and Environmental Sciences, https://www.esci.umn.edu/facilities/pillsbury.
39. Alexander Winchell to Caroline Winchell, 11 December 1890, box 3, folder correspondence 1885–90, Newton Horace Winchell and Family Papers.

15. *THE AMERICAN GEOLOGIST*

1. "Editorial Comment," *American Geologist* 10, no. 6 (December 1892): 385. Unfortunately, *The American Geologist*'s editorials were never signed.
2. H. Foster Bains, "N. H. Winchell and the American Geologist," *Economic Geology* 11 (1916): 60.
3. "Personal and Scientific News," *American Geologist* 1, no. 1 (January 1888): 67–68.
4. "Introductory," *American Geologist* 1, no. 1 (January 1888): 1–3.
5. "Editorial Comment," *American Geologist* 10, no. 6 (December 1892): 387.
6. Newton H. Winchell to Alexander Winchell, 31 May 1888, box 2, folder 64, Alexander Winchell Papers, Bentley Historical Library, University of Michigan, Ann Arbor.
7. Newton H. Winchell to Alexander Winchell, 19 June 1888.
8. Newton H. Winchell to Alexander Winchell, 16 April 1888.
9. A. Winchell, "The Unconformities of the Animike in Minnesota," *American Geologist* 1, no. 1 (January 1888): 14–24.
10. [A. Winchell], "Geology in the Educational Struggle for Existence," *American Geologist* 1, no. 1 (January 1888): 36–44. This editorial was unattributed, but Bains in "N. H. Winchell and the American Geologist," page 59, states it came off A. Winchell's pen.
11. Horace Vaughn Winchell, "A Bit of Iron Range History," *American Geologist* 13, no. 3 (March 1894): 164.
12. "Opening of the Mesabi Iron Range," MNopedia, mnopedia.org/event/opening-mesabi-iron-range.
13. Warren Upham, "Causes and Conditions of Glaciation," *American Geologist* 14, no. 1 (July 1894): 12–20.
14. Warren Upham, "Causes, Stages, and the Time of the Ice Age," *Popular Science Monthly* 49 (July 1896), https://en.wikisource.org/wiki/Popular_Science_Monthly/Volume_49/July_1896/Causes,_Stages,_and_the_Time_of_the_Ice_Age.
15. Richard Ojakangas, personal communication with author, 7 September 2017.
16. Upham, "Causes, Stages, and the Time of the Ice Age."
17. "Review of Recent Literature," *American Geologist* 1, no. 1 (January 1888): 61.
18. "Personal and Scientific News," *American Geologist* 12, no. 1 (July 1893): 65.
19. "Review of Recent Geological Literature," *American Geologist* 13, no. 3 (March 1894): 194.
20. Bains, "N. H. Winchell and the American Geologist," 60.

21. F. B. Taylor, "A Reconnaissance of Abandoned Shorelines of the South Coast of Lake Superior," *American Geologist* 13, no. 6 (June 1984): 365–83.
22. "Continental Drift," The Emergence and Evolution of Plate Tectonics, publish .illinois.edu, publish.illinois.edu/platetechtonics/continental-drift/.
23. Encyclopedia.com, s.v. "Taylor, Frank Bursley," https://www.encyclopedia.com /people/science-and-technology/geology-and-oceanography-biographies /frank-bursley-taylor.
24. Thomas S. Kuhn, *The Structure of Scientific Revolutions,* 2d ed. (Chicago: University of Chicago Press, 1970), 144.
25. Louise Winchell Dayton Denman, undated (but probably 1941), "Life with Father," envelope 2, Newton Horace Winchell and Alexander N. Winchell Papers, Biography Files, Hennepin County Library Special Collections. Also H. A. Beaudoux, untitled, sent as a letter to Avis Winchell Grant, 1941, envelope 2.

16. TERMINAL MORAINE

1. H. A. Beaudoux to Avis Winchell Grant, 23 January 1941, envelope 2, Newton Horace Winchell and Alexander N. Winchell Papers, Biography Files, Hennepin County Library Special Collections.
2. Minutes, 3 May 1892, University of Minnesota Board of Regents, 9–10, conservancy.umn.edu//handle/11299/48931.
3. A geodetic survey is one that takes into account the curvature of the earth.
4. Minutes, 19 July 1893, University of Minnesota Board of Regents, 59, conservancy.umn.edu/bitstream/handle/11299/48931/Regents %201889–1895.pdf?sequence=1&isAllowed=y.
5. N. H. Winchell, "Address," *The Minnesota Geological and Natural History Survey: The Twentieth Annual Report for the Year 1891* (Minneapolis: Harrison and Smith, 1893), iv.
6. Grant's thesis topic for his doctorate, taken at Johns Hopkins University under George H. Williams, was on the geology of Kekaquabic Lake on the eastern end of the Mesabi Range.
7. Winchell, *The Minnesota Geological and Natural History Survey: Twentieth Annual Report,* 114.
8. N. H. Winchell, *The Minnesota Geological and Natural History Survey: The Twenty-First Annual Report for the Year 1892* (Minneapolis: Harrison and Smith, 1893), 128.
9. Ibid., 142.
10. H. A. Beaudoux to Avis Winchell Grant, 23 January 1941.
11. N. H. Winchell, *The Geological and Natural History Survey of Minnesota: The Twenty-Third Annual Report for the Year 1894* (Minneapolis: Harrison and Smith, 1895), 5.
12. For example, see Minutes, 22 December 1894, University of Minnesota Board of Regents, conservancy.umn.edu/bitstream/handle/11299/48931/Regents %201889–1895.pdf?sequence=1&isAllowed=y.
13. "Home Hits and Happenings/How We Spent the Summer," *Ariel* 15, no. 112 (September 1891), 8.

14. H. A. Beaudoux to Avis Winchell Grant, 23 January 1941.
15. *Minneapolis Tribune*, 4 March 1895.
16. The *Women's Journal* was a women's rights periodical founded by Lucy Stone and her husband, Henry Blackwell.
17. All quotes from (illegible) Minneapolis newspaper clipping, 22 November 1893; box 3, folder correspondence 1893–94, Newton Horace Winchell and Family Papers, Minnesota Historical Society (MHS), St. Paul.
18. Undated, unsourced newspaper clipping, box 3, correspondence 1893–94, Newton Horace Winchell and Family Papers.
19. Numerous handbills from Ida Belle Winchell's performances, box 3, folder 1891–92, Newton Horace Winchell and Family Papers, MHS.
20. H. A. Beaudoux to Avis Winchell Grant, 23 January 1941.
21. Louise Winchell Dayton Denman, undated (but probably 1941), "Life with Father," envelope 2, Newton Horace Winchell and Alexander N. Winchell Papers, Biography Files, Hennepin County Library Special Collections.

17. AMERICANS IN PARIS

1. Newton Horace Winchell, journal, 30 September 1857, box 12, vol. 10, Newton Horace Winchell and Family Papers, Minnesota Historical Society (MHS), St. Paul.
2. Walter Summers to Mrs. C. A. Summers, 22 March 1890, "A Letter from S.S. *City of Paris*," What Would the Founders Think?, whatwouldthefoundersthink .com/a-letter-from-the-city-of-paris.
3. "City of Paris," TheShipsList, http://www.theshipslist.com/ships/descriptions /ShipsCC.shtml.
4. Wikipedia, s.v. "Paris in the Belle Époque," en.wikipedia.org/wiki/Paris_in _the_Belle Epoque.
5. Mary Sperling McAuliffe, *Dawn of the Belle Epoque: The Paris of Monet, Zola, Bernhardt, Eiffel, Debussy, Clemenceau, and Their Friends* (Lanham, Md.: Rowman and Littlefield, 2011), 167, 188.
6. Ibid., 196.
7. Encyclopaedia Britannica, s.v. "Alfred Lacroix," https://www.britannica.com /biography/Alfred-Lacroix.
8. A. Lacroix, "Les enclaves des roches volcaniques," *Annales de l'Académie de Mâcon* 10 (1893).
9. Sibylle Franks and Rudolf Trümpy, "The Sixth International Geological Congress: Zürich 1894," *Episodes* 28, no. 3 (2005): 189.
10. N. H. Winchell, "Microscopic Characters of the Fisher Meteorite (Minnesota No. 1)," *American Geologist* 17, no. 3 (March 1896): 173–76.
11. N. H. Winchell, *The Geological and Natural History Survey of Minnesota: The Twenty-Fourth (and Final) Annual Report for the Years 1895–1898* (Minneapolis: University Press, 1899), vii–xxviii.
12. Charlotte S. Winchell, "World's WCTU Convention," *Minnesota White Ribboner* 5, no. 19 (1895): 6–8.
13. Ibid., 7.
14. Charlotte S. Winchell, "A Few Observations upon the Custom and Effect of

Wine Drinking in France," *White Ribboner* 6, no. 1 (1895), box 3, folder 1895–96, Newton Horace Winchell and Family Papers.

15. Ibid.
16. Alexander Winchell, N. H. Winchell, and Alexander Newton Winchell, "Louise #3907," *The Winchell Genealogy: The Ancestry and Children of Those Born to the Winchell Name in America since 1935,* 2d. ed. (Minneapolis: H. V. Winchell, 1917), 440.
17. Charlotte S. Winchell to Alexander N. Winchell, 9 February 1896, box 3, folder 1895–96, Newton Horace Winchell and Family Papers.
18. French women got the right to vote in 1945.
19. Charlotte S. Winchell to Alexander N. Winchell, 9 February 1896. Carrara marble, from Italy, was favored by sculptors Michelangelo and Bernini.
20. Ibid.
21. Ibid.
22. Charlotte S. Winchell to Alexander N. Winchell, 22 March 1896, box 3, folder 1895–96, Newton Horace Winchell and Family Papers.
23. Probably gabbro or a dark diorite; Richard Ojakangas, email to author, 4 December 2017.
24. Charlotte S. Winchell to Alexander N. Winchell, 22 March 1896.
25. Ibid.
26. Ibid.
27. Charlotte S. Winchell to Alexander N. Winchell, 19 April 1896, box 3, folder 1895–96, Newton Horace Winchell and Family Papers.
28. Ibid.

18. GLACIAL RETREAT

1. Newton Horace Winchell to the Board of Regents of the University of Minnesota, 10 December 1897, box 3, folder correspondence 1896–97, Newton Horace Winchell and Family Papers, Minnesota Historical Society (MHS), St. Paul, 1.
2. Richard W. Ojakangas and Charles L. Matsch, *Minnesota's Geology* (Minneapolis: University of Minnesota Press, 1982), 148. However, in 2018, the Ontario gold district northwest of Fort Frances was gearing up for operation.
3. Winchell to the Board of Regents, 10 December 1897, 3.
4. Ibid., 5.
5. Ibid., 7.
6. Ibid.
7. Ibid., 8
8. Ibid., 9.
9. Minutes, 12 April 1898, University of Minnesota Board of Regents, 189–90, conservancy.umn.edu/handle/11299/48932.
10. Ida Belle Winchell to Horace V. Winchell, 26 May 1898, box 4, folder correspondence May–September 1898, Newton Horace Winchell and Family Papers.
11. Ida Belle Winchell to Horace V. Winchell, 28 May 1898, box 4, folder correspondence May–September 1898.

12. Ida Belle Winchell to Horace V. Winchell, 30 May 1898, box 4, folder correspondence May–September 1898.
13. Ida Belle Winchell to Horace V. Winchell, 2 June 1898, box 4, folder correspondence May–September 1898.
14. Ida Belle Winchell to Horace V. Winchell, 13 June 1898, box 4, folder correspondence May–September 1898.
15. C. L. Herrick to the Board of Regents of the University of Minnesota, 23 November 1899, box 4, folder correspondence March–December 1899, Newton Horace Winchell and Family Papers.
16. John J. Stevenson to Newton H. Winchell, 19 September 1898, box 4, folder correspondence May–September 1898, Newton Horace Winchell and Family Papers.
17. Herrick to the Board of Regents, 23 November 1899.
18. *Minneapolis Tribune,* "The Geological Survey," 9 September 1899, clipping, box 4, folder correspondence March–December 1899, Newton Horace Winchell and Family Papers.
19. *Minneapolis Journal,* 13 September 1899, clipping, box 4, folder correspondence March–December 1899, Newton Horace Winchell and Family Papers.
20. J. B. Woodworth to Newton H. Winchell, 30 September 1898, box 4, folder correspondence May–September 1898, Newton Horace Winchell and Family Papers.
21. Nathaniel S. Shaler to Newton H. Winchell, 13 September 1898, box 4, folder correspondence May–September 1898.
22. Horace V. Winchell to Newton H. Winchell, 29 October 1898, box 4 folder correspondence October–December 1898.
23. S. R. Winchell to Newton H. Winchell, 6 December 1898, box 4 folder correspondence October–December 1898.
24. Lucius Hubbard to Newton H. Winchell, 16 February 1899, box 4 folder correspondence January–February 1899.
25. Marshman E. Wadsworth to Newton H. Winchell, 22 February 1899, box 4, folder correspondence January–February 1899.
26. S. R. Winchell to Newton H. Winchell, 23 March 1899, box 4, folder correspondence March–December 1899.
27. Israel C. Russell to Newton H. Winchell, 27 March 1899, box 4, folder correspondence March–December 1899.
28. Victor W. Lyon (city engineer for Jeffersonville, Indiana) to Newton H. Winchell, 15 July 1900, box 4, folder correspondence 1900.
29. Charles Schuchert to Newton H. Winchell, 27 August 1900, box 4, folder correspondence 1900.
30. Warren Upham to Newton H. Winchell, 15 October 1900, box 4, folder correspondence 1900.
31. A. F. Bechtold to Newton H. Winchell, 10 November 1900, box 4, folder correspondence 1900.
32. Newton H. Winchell to John S. Pillsbury, 5 December 1898, box 4, folder correspondence October–December 1898.
33. I. White to Newton H. Winchell, undated July 1901, box 5, folder correspondence 1901, Newton Horace Winchell and Family Papers.

34. M. S. Huston to Charlotte I. Winchell, 8 July 1901, box 5, folder correspondence 1901.
35. D. Draper Dayton to E. W. Dayton, undated, 1901, box 1, folder 1, George D. Dayton II and Family Papers, MHS.
36. Horace V. Winchell to Newton H. Winchell, 8 July 1901, box 5, folder correspondence 1901, Newton Horace Winchell and Family Papers.
37. "A History of 100 Degrees in the Twin Cities," Minnesota Department of Natural Resources, https://www.dnr.state.mn.us/climate/journal/100degreesmsp.html.
38. A. P. Coleman to Newton H. Winchell, 26 November 1901, box 5, folder correspondence 1901.
39. Ida Belle Winchell to Horace V. Winchell, 26 May 1898, box 4, folder correspondence May–September 1898.
40. Newton H. Winchell to Charlotte I. Winchell, 11 May 1903, box 5, folder correspondence 1902–5, Newton Horace Winchell and Family Papers.
41. Ibid.
42. Newton H. Winchell to Charlotte I. Winchell, 19 May 1903, box 5, folder correspondence 1902–5.
43. Newton H. Winchell to Charlotte I. Winchell, 31 May 1903, box 5, folder correspondence 1902–5.
44. Ibid.
45. Newton H. Winchell to Charlotte I. Winchell, 3 September 1903, box 5, folder correspondence 1902–5.
46. Ibid.
47. N. H. Winchell, "Review of Recent Geological Literature: *The Planetary System* by Frank Taylor," *American Geologist* 33, no. 3 (March 1904): 191.
48. Warren Upham to Newton H. Winchell, 4 June 1904, box 5, folder correspondence 1902–5.
49. Horace V. Winchell to Charlotte I. Winchell, 2 March 1905, box 5, folder correspondence 1902–5.
50. Charles Mayo to Alexander N. Winchell, 11 March 1905, box 5, folder correspondence 1902–5.
51. Ibid.
52. Alexander N. Winchell to Newton H. Winchell, 14 March 1905, box 5, folder correspondence 1902–5.

19. THE ARCHAEOLOGIST

1. Helen Clapesattle, *The Brothers Mayo* (Rochester, Minn.: Mayo Foundation for Medical Education and Research, 1969), 232.
2. N. H. Winchell, "Editorial Comment," *American Geologist* 36, no. 4 (October 1905): 257.
3. Wikipedia, s.v. "Willamette Meteorite," https://en.wikipedia.org/wiki/Willamette_Meteorite.
4. Winchell, "Editorial Comment," 257.
5. Horace V. Winchell to owner of *Economic Geology,* 28 July 1905, box 5, folder

correspondence 1902–5, Newton Horace Winchell and Family Papers, Minnesota Historical Society (MHS), St. Paul.

6. Ulysses S. Grant to Newton H. Winchell, 5 August 1905, box 5, folder correspondence 1902–5.
7. Oliver P. Hay to Newton H. Winchell, July 1905, box 5, folder correspondence 1902–5.
8. Ulysses S. Grant to Newton H. Winchell, 5 August 1905, box 5, folder correspondence 1902–5.
9. Florence Bascom to Newton H. Winchell, 10 July 1905, box 5, folder correspondence 1902–5.
10. Horace V. Winchell to owner of *Economic Geology*, 28 July 1905.
11. Horace V. Winchell to Newton H. Winchell, 21 August 1905, box 5, folder correspondence 1902–5.
12. Horace V. Winchell to Newton H. Winchell, September 1905, box 5, folder correspondence 1902–5.
13. N. H. Winchell, "Editorial Comment," *American Geologist* 36, no. 5 (November 1905): 311.
14. Ibid., 321–22.
15. Warren Upham, Report of the Museum Committee of January 26, 1906, to the Minnesota History Society Council meeting, February 12, 1906, folder minutes and reports of the Minnesota Historical Society, 1904–6, box locator 303.G.14.7B, Warren Upham Papers, MHS Archives.
16. Jacob Brower, 7 May 1903, vol. 82, box locator 303.G.13.7B, Jacob V. Brower Papers, MHS Archives.
17. Brower, 20 August 1903, vol. 81, Jacob V. Brower Papers; see vol. 80 for offer as go-between.
18. Brower, 12 May 1903, vol. 82, Jacob V. Brower Papers.
19. John Sanborn, presidential address to the Minnesota Historical Society, 11 January 1904, folder MHS paper, minutes, and reports of the Minnesota Historical Society, 1904–6, box locator 303.G.14.7B, Warren Upham Papers, MHS Archives.
20. Compare to the New England state libraries in 1905: Maine, 119 volumes; New Hampshire, 148 volumes; Vermont, 45 volumes; Massachusetts, 690 volumes; Rhode Island, 55 volumes; and Connecticut, 152 volumes.
21. Third site of the Science Museum of Minnesota.
22. Aaron Isaacs, coauthor of *Twin Cities by Trolley: The Streetcar Era in Minneapolis and St. Paul,* email to author.
23. Larry Millett, *Lost Twin Cities* (St. Paul: Minnesota Historical Society Press, 1992), 190–91.
24. George Rapp, "Prologue: The Organization, Development, and Future of Geoarchaeology," in *Reconstructing Human-Landscape Interactions,* ed. Lucy Wilson, Pam Dickinson, and Jason Jeandron (Newcastle, UK: Cambridge Scholars Publishing, 2007).
25. See, for example, N. H. Winchell, *The Geological and Natural History Survey of Minnesota: The Fifth Annual Report for the Year 1876* (St. Paul: Pioneer Press Co., 1877), 50.

26. George H. Squier to Newton H. Winchell, 7 November 1908, folder correspondence 1908–9, Winchell Correspondence, box locator 303.G.12.9B, MHS Archives.
27. L. A. Ogaard to Newton H. Winchell, 9 December 1907, folder correspondence 1906–7, Winchell Correspondence, box locator 303.G.12.9B, MHS Archives.
28. "Archaeology in Trempealeau, Wisconsin," Mississippi Valley Archaeology Center at the University of Wisconsin-La Crosse, http://mvac.uwlax.edu/past-cultures/specific-sites/trempeleau/.
29. Thomas Hughes to Newton H. Winchell, 3 September 1909, folder correspondence 1908–9, Winchell Correspondence, box locator 303.G.12.9B, MHS Archives.
30. Thomas Hughes to Newton H. Winchell, 8 October 1910, folder correspondence 1908–9.
31. Ian Dyck, "The Life and Work of W. B. Nickerson," YouTube video, 2017, https://www.youtube.com/watch?v=2s5Jyus1OYs.
32. William Baker Nickerson to Newton H. Winchell, 20 July 1913, folder correspondence June–October 1913, Winchell Correspondence, box locator 303.G.12.9B, MHS Archives.
33. George R. Holley and Michael G. Michlovic, *The Prehistoric Village Cultures of Southern Minnesota* (Moorhead: Minnesota State University Moorhead, 2013), mn.gov/admin/assets/2013-Prehistoric-Village-Cultures-of-Southern-Minnesota.
34. Ibid., 20.
35. Joseph Buisson to Newton H. Winchell, 24 January 1910, folder correspondence ca. 1910–11, Winchell Correspondence, box locator 303.G.12.9B, MHS Archives.
36. C. H. Beaulieu to Newton H. Winchell, 2 July 1910, folder correspondence ca. 1910–11.
37. "History: The First Archaeologists," Minnesota Department of Administration, State Archaeologist, mn.gov.admin/archaeologist/the-public/research/history/.
38. Joe Alan Artz, Emilia L. D. Bristow, and William E. Whittaker, *Mapping Precontact Burial Mounds in Sixteen Minnesota Counties Using Light Detection and Ranging (LiDAR),* Contract Completion Report 1976 (Iowa City: Office of the State Archaeologist, 2013), 2.
39. Translation by Erik Wahlstrom, professor of Scandinavian languages, University of California, Los Angeles, 1958, quoted in Theodore C. Blegen, *The Kensington Rune Stone: New Light on an Old Riddle* (St. Paul: Minnesota Historical Society Press, 1968), 11.
40. Blegen, *The Kensington Rune Stone,* 53.
41. Herman, Minnesota, is twenty-three miles west of Kensington.
42. Hjalmar R. Holand to Newton H. Winchell, 3 August 1908, folder correspondence with Winchell and Other Papers, 1906–March 1910, Kensington Rune Stone, Archaeological Records, box locator 303.G.12.9B, MHS Archives.
43. Blegen, *The Kensington Rune Stone,* 58.
44. Ibid.

45. Ibid., 54.
46. Hjalmar R. Holand to Newton H. Winchell, 3 August 1908.
47. Blegen, *The Kensington Rune Stone,* 72–79.
48. Ibid., 80–81.
49. December 17, 1909, agreement between Hjalmar R. Holand and the Minnesota Historical Society, folder correspondence with Winchell and Other Papers, 1906–March 1910, Kensington Rune Stone, Archaeological Records, box locator 303.G.12.9B, MHS Archives.
50. Hjalmar R. Holand to Newton H. Winchell, 11 January 1910, folder correspondence with Winchell and Other Papers, 1906–March 1910.
51. George O. Curme to Newton H. Winchell, 9 March 1910, folder correspondence with Winchell and Other Papers, 1906–March 1910.
52. C. N. Gould to Warren Upham, 19 March 1910, folder correspondence with Winchell and Other Papers, 1906–March 1910.
53. Ibid.
54. N. H. Winchell, "Report on the Kensington Rune Stone," April 1910, box locator 303.G.12.9B, MHS Archives, 48.
55. Winchell, "Report on the Kensington Rune Stone."
56. Blegen, *The Kensington Rune Stone,* 90–91.
57. Ibid., 13.
58. Winchell, "Report on the Kensington Rune Stone," 68–69.
59. J. R. Swanton to Newton H. Winchell, 10 May 1910, folder correspondence 1908, Winchell Correspondence, box locator 303.F. 13.8F, MHS Archives.
60. "Steamer Goes Down Off Alaska; 5 Lost," *Minneapolis Tribune,* 28 August 1909, 1.
61. Newton H. Winchell, *The Aborigines of Minnesota* (St. Paul: Minnesota Historical Society, 1911), 559.
62. Warren K. Moorehead to Newton H. Winchell, 7 October 1911, folder 1910–1911, Winchell Papers, box locator 303.G.12.9B, MHS Archives.
63. William Watts Folwell to Newton H. Winchell, 29 September 1911, folder Winchell Correspondence 1911–12, box locator 307.F.13.9B, MHS Archives.
64. Warren Upham to Newton H. Winchell, 15 September 1911, folder Winchell Correspondence 1911–12.

20. WINCHELL THE ROCK

1. N. H. Winchell, *The Geological and Natural History Survey of Minnesota: The Sixth Annual Report for 1877* (Minneapolis: Johnson, Smith and Harrison, 1878), 53–56.
2. Newton H. Winchell, *The Aborigines of Minnesota* (St. Paul: Minnesota Historical Society Press, 1911), 17, 18.
3. Hilary L. Chester, "Frances Eliza Babbitt and the Growth of Professionalism of Women in Archaeology," in *New Perspectives on the Origins of Americanist Archaeology,* ed. David L. Browman and Stephen Williams (Tuscaloosa: University of Alabama Press, 2002), 169.

4. Warren Upham, "Man in the Ice Age at Lansing, Kansas, and Little Falls, Minnesota," *American Geologist* 30, no. 3 (September 1902): 147–48.
5. Later C-14 dating put the skeleton at 5,500 years old. Wikipedia, s.v. "Lansing Man," https://en.wikipedia.org/wiki/Lansing_Man.
6. Upham, "Man in the Ice Age," 150.
7. For a discussion of this and other theories, see "Humans and the Extinction of Megafauna in the Americas," *Dartmouth Undergraduate Journal of Science,* spring 2009, http://dujs.dartmouth.edu/2009/05/humans-and-the-extinction-of-megafauna-in-the-americas/.
8. N. H. Winchell, "Was Man in America in the Glacial Period?," *Bulletin of the Geological Society of America* 14 (1903): 138, 139.
9. Warren Upham, "The Work of N. H. Winchell in Glacial Geology and Archaeology," *Economic Geology* 11, no. 1 (1916): 63–72.
10. Frederick W. Hodge to Newton H. Winchell, 22 June 1912, folder 1912–March 5, 1913, box locator 303.G.12.9B, Newton H. Winchell Papers, Minnesota Historical Society (MHS) Archives, St. Paul.
11. Newton H. Winchell to Frederick W. Hodge, 24 June 1912, folder 1912–March 5, 1913.
12. N. H. Winchell, *The Weathering of Aboriginal Stone Artifacts* (St. Paul: Minnesota Historical Society Press, 1913), vii.
13. Chester, "Frances Eliza Babbitt," 170.
14. Ibid., 171–72.
15. David Browman, "Frederic Ward Putnam: Contributions to the Development of Archaeological Institutions and Encouragement of Women Practitioners," in *New Perspectives on the Origins of Americanist Archaeology,* ed. Browman and Williams, 217.
16. Chester, "Frances Eliza Babbitt," 172.
17. Newton H. Winchell to Charles A. Abbott, 19 November 1913, folder correspondence November 1913–May 1914, box locator 303.G.12.9B, Winchell Papers, MHS Archives.
18. W. K. Moorehead, "A Review of 'The Weathering of Aboriginal Stone Artifacts,'" 1913, folder correspondence June–October 1813, box locator 303.G.12.9B, Winchell Papers, MHS Archives.
19. D. V. Daigneau to Newton H. Winchell, 23 August 1913, folder correspondence June–October 1913.
20. John J. Stevenson to Warren Upham, May 1914, Winchell Incoming Correspondence 1913–14, box locator 307.F.13.9B, MHS Archives.
21. J. B. Woodworth to Warren Upham, 12 May 1914, Winchell Incoming Correspondence 1913–1914, MHS, box locator 307.F.13.9B, MHS Archives.
22. Ernest Volk to Newton H. Winchell, 29 January 1914, folder correspondence November 1913–May 1914, box locator 303.G.12.9B, Winchell Papers, MHS Archives.
23. Newton H. Winchell to Warren Upham, 25 March 1914, November 1913–May 1914.
24. M. L. Arey to Charlotte S. Winchell, 13 May 1914, Winchell Incoming Correspondence 1913–1914.

25. Charlotte S. Winchell, (undated, but 1914), "A Record of the Symptoms, Sickness and Treatment of Our Family in Every Case of Illness Occurring after Marriage August 24, 1864," box 13, vol. 14, Newton Horace Winchell and Family Papers, MHS.

EPILOGUE

1. W. W. Folwell, 1914, folder Winchell 1914, box locator 307.F.13.9B, Minnesota Historical Society (MHS) Archives, St. Paul.
2. J. Frank Stout, 1914, folder Winchell 1914.
3. George H. Stone to Warren Upham, 30 May 1914, folder Winchell 1914.
4. Edward S. Dana to Warren Upham, 21 May 1914, folder Winchell 1914.
5. John J. Stevenson to Warren Upham, undated, folder Winchell 1914.
6. Warren Upham, "Memoir of Newton Horace Winchell," *Bulletin of the Geological Society of America* 26 (1915): 27–46; Warren Upham, "The Work of N. H. Winchell in Glacial Geology and Archaeology," *Economic Geology* 11, no. 1 (1916): 63–72. Warren Upham, "Newton Horace Winchell, 1839–1914," *Bulletin of the Minnesota Academy of Science* 5, no. 2 (1914): 78–88.
7. Upham, "Memoir of Newton Horace Winchell," 31.
8. "Winchell Plan for Woman Students' Home Realized at the University of Minnesota after 43 Years," *Minneapolis Journal,* 11 July 1915, newspaper clipping, Newton Horace Winchell (1839–1914), Charlotte I. (1836–1926) and Family, Hennepin County Library Special Collections.
9. Horace V. Winchell, "Keep Eyes on Russia Money Out of It," 22 September 1917, newspaper clipping, Horace V. Winchell (1862–1923), Biography Files, Hennepin County Libraries Special Collections.
10. David Backes, *A Wilderness Within: The Life of Sigurd F. Olson* (Minneapolis: University of Minnesota Press, 1999), 47.
11. Warren Upham, *Minnesota Place Names: A Geographical Encyclopedia,* 3rd. ed. (St. Paul: Minnesota Historical Society Press, 2001), 144, 315, 314, 145, 541.
12. George M. Schwartz, "Newton Horace Winchell: A Tribute," 2012, University of Minnesota, Department of Earth and Environmental Sciences, http://www.esci.umn.edu/winchell.
13. L. H. Thorleifson, undated "Review of Lake Agassiz History," https://www.esci.umn.edu/sites/www.esci.umn.edu/files/review%20of%20lake%20agassiz%20history.pdf.
14. F. Garvin Davenport, "Newton H. Winchell, Pioneer of Science," *Minnesota History* 32, no. 4 (December 1951): 222.
15. E. W. Davis, "Taconite: The Derivation of the Name," *Minnesota History* 33, no. 7 (fall 1953): 282.
16. George Rapp, "Prologue: The Organization, Development, and Future of Geoarchaeology," in *Reconstructing Human-Landscape Interactions,* ed. Lucy Wilson, Pam Dickinson, and Jason Jeandron (Newcastle, UK: Cambridge Scholars Publishing, 2007).

INDEX

Sue Leaf is the author of *Potato City: Nature, History, and Community in the Age of Sprawl* and the Midwest Book Award winner *Portage: A Family, a Canoe, and the Search for the Good Life* (Minnesota, 2015). Her books *The Bullhead Queen: A Year on Pioneer Lake* and *A Love Affair with Birds: The Life of Thomas Sadler Roberts,* both published by the University of Minnesota Press, were finalists for the Minnesota Book Award.